WATER RESOURCE PLANNING, DEVELOPMENT AND MANAGEMENT

PRINCIPLES AND PRACTICES OF WATER RESOURCES DEVELOPMENT AND MANAGEMENT

WATER RESOURCE PLANNING, DEVELOPMENT AND MANAGEMENT

Additional books in this series can be found on Nova's website
under the Series tab.

Additional e-books in this series can be found on Nova's website
under the e-books tab.

WATER RESOURCE PLANNING, DEVELOPMENT AND MANAGEMENT

PRINCIPLES AND PRACTICES OF WATER RESOURCES DEVELOPMENT AND MANAGEMENT

M. HOSSAIN ALI

nova publishers
New York

Library of Congress Cataloging-in-Publication Data

ISBN: 978-1-63485-175-6

Names: Ali, M. H. (Md. Hossain), author.
Title: Principles and practices of water resources development and management / Hossain Ali (Agricultural Engineering Division, Bangladesh Institute of Nuclear Agriculture (BINA), Principal Scientific Officer, and Head, Bangladesh).
Description: Hauppauge, New York : Nova Science Publisher's, Inc., 2016. | Series: Water resource planning, development and management | Includes bibliographical references and index.
Identifiers: LCCN 2016017041 (print) | LCCN 2016017569 (ebook) | ISBN 9781634851756 (hardcover) | ISBN 9781634852050 ()
Subjects: LCSH: Water resources development.
Classification: LCC TC405 .A59 2016 (print) | LCC TC405 (ebook) | DDC 333.79--dc23
LC record available at https://lccn.loc.gov/2016017041

Published by Nova Science Publishers, Inc. † New York

CONTENTS

PREFACE

Freshwater management challenges are increasingly common. Allocation of limited water resources between agricultural, municipal and environmental uses now requires the full integration of supply, demand, water quality and ecological considerations.

Water is the scarcest resource. The importance of the resource for the survival of the modern society - sustaining agricultural and industrial growth, and the retardation of environmental degradation needs no elaboration. Sustainable development and management of the resource require scientific and systematic approaches. The book *Principles and Practices of Water Resources Development and Management* is designed to cover the major aspects of water resources development and management such as assessment of the resources, estimation of groundwater recharge, water-well construction and groundwater hydraulics, management of the resources, water contamination, protection of the resources, economics in water resources, statistical methods in water resources, and use of models in water resources management. When necessary, workout problems are provided to explain the application of theory/methodology in practice. This comprehensive and compact presentation of the book will serve as a textbook for undergraduate students in Civil Engineering, Environmental Engineering, Agricultural Engineering, Water Resources Engineering, and Geotechnical/Geo-science Engineering. Students of other relevant branches such as Hydrology, Geology, Hydrogeology, Geochemistry, Bio-Science Engineering, and engineers working in the field and at research institutes will also be benefited from lessons within its pages. Although the target audience of the book is undergraduate students, post-graduate students will also learn from this book. Considering the topics and depth covered, engineers, scientists, practitioners and educators will find this book as a valuable resource.

Chapter 1

INTRODUCTION – PERSPECTIVES OF WATER RESOURCES MANAGEMENT

ABSTRACT

Water resources of a country constitute one of its vital assets. Of all the planet's renewable resources, water has a unique place. It is essential for sustaining all forms of life, food production, economic development, and for general wellbeing of human. It is impossible to substitute for most of its uses, difficult to de-pollute, expensive to transport, and it is truly a unique gift to mankind from nature. Water is also one of the most manageable of the natural resources as it is capable of diversion, transport, storage, and recycling. All these properties impart to water its great utility for human beings. Today, in most countries suffering water shortages, at the heart is the question of whether a water crisis can be averted or whether water can be made productive.

1.1. WATER AND THE SOCIETY

Water has been identified as a crucial resource for all life, production, and development, while a lack of access to water has been linked to poverty. It is desired to use water in more efficient, equitable, and environmentally sustainable ways. These different links and objectives touch upon many disciplines and people working in and/or dependent on water: economists, sociologists, engineers, politicians, decision-makers, and other stakeholders.

The surface water and groundwater resources of a country play a major role in agriculture, hydropower generation, livestock production, industrial activities, fisheries, navigation, recreational activities, etc.

1.2. AVAILABILITY OF WATER ON THE EARTH

Over 70% of our Earth's surface is covered by water. Although water is seemingly abundant, the real issue is the amount of fresh water available. About 97.5% of all water on Earth is salt water, leaving only 2.5% as fresh water. Nearly 70% of that fresh water is frozen in the icecaps of Antarctica and Greenland; most of the remainder is present as soil moisture, or lies in deep underground aquifers as groundwater not accessible to human use. Only ~1% of the world's fresh water is accessible for direct human uses. This is the water found in lakes, rivers, reservoirs and those underground sources that are shallow enough to be tapped at an affordable cost. Only this amount is regularly renewed by rain and snowfall, and is therefore available on a sustainable basis.

A great deal of water use is non-consumptive, which means that the water is returned to surface runoff. Usually that water is contaminated however, whether used for agriculture, domestic consumption, or industry.

1.3. Hydrologic Cycle and Human Intervention

Water circulates naturally. It moves between the air, the land, underground, to the ocean and back again via the water cycle. When it evaporates, it changes from liquid to gas and eventually recondenses as a liquid. Water assimilated during photosynthesis becomes part of carbohydrates stored in plants, but ultimately reverts to water again by decomposition.

When used, water loses properties such as purity, heat content, and potential gravitational energy, but eventually, most degraded water resources are refreshed by natural processes in the hydrological cycle, which is mostly driven by solar energy. When considering water flux as the most relevant measure of water resources, the speed of water circulation becomes crucial. Mean residence times of water molecules—i.e., how long they stay in a given reservoir—can be estimated by dividing the volume of the reservoir by the mean flux into and out of it. For rivers unaffected by human interventions, the mean residential time of water is about two and a half weeks. In contrast, the recharge rate of some groundwater aquifers is very slow, and the mean residential time is considered to be hundreds or even thousands of years.

The hydrologic cycle is being modified quantitatively and qualitatively in most river basins as a result of the development activities such as construction of dams and reservoirs, barrage, irrigation, land-use change, etc. Historically, society's response to floods and droughts has been to impound surface water in reservoirs, and to release it as needed. However, a dearth of geologically suitable locations for new dams, combined with increased awareness of their ecological consequences, will hinder this response to future hydrologic extremes, even as their frequency and intensity increase. Long-term temperature rises will increase the need to store water for distribution over a longer dry season. In some areas, an integrated solution can be achieved by artificially recharging excess runoff, when available. Thus, depleted aquifers can be transformed into underground "reservoirs" to supplement the flood- and drought-buffering capacity of existing surface-water reservoirs.

1.4. Global Climate Change and the Water Resource

Global climate change will profoundly affect hydrologic systems worldwide. Glacial melting and increasing ocean temperatures lead to sea-level rise. On the continents, the frequency and severity of floods and droughts are expected to increase, while higher temperatures will reduce winter snowpack and hasten spring snowmelt from mountainous areas. Higher temperature will increase water demand in agricultural sector (Ali and Adham, 2007). Unchecked, groundwater depletion can exacerbate the impacts of these changes; conversely, controlled management of groundwater depletion can contribute to their mitigation.

Assuming that the volume of groundwater depleted during the past 100 years is much greater than can be accounted for by nontransient increases in volumes of water stored in soil, natural channels and lakes, or the atmosphere, then the ultimate sink for the "missing" groundwater is the oceans. Worldwide, the magnitude of groundwater depletion from storage may be so large as to constitute a measurable contributor to sea-level rise. For example, the total volume depleted from the High Plains aquifer equates to about 0.75 mm, or about 0.5%, of the observed sea-level rise during the 20th century. Reducing future groundwater depletion (and increasing groundwater storage) can help in a small way to reduce future sea-level rise.

1.5. Future Challenges of Water Resources Management

Managers of both surface and groundwater are facing new challenges of fulfilling not only the traditional objectives of securing water supplies, but also of improving and protecting ecological health, while facing greater climatic fluctuations and population pressure. To achieve consensus, managers must balance the competing needs of people, industry, agriculture, and the environment.

In developing countries, where the livelihoods of millions of poor people may depend on unsustainable groundwater withdrawals, water managers face additional complexities that are not amenable to engineering solutions alone. In the future, the pressure of increasing populations worldwide may foster greater acceptance of groundwater depletion, regardless of a nation's development stage.

To minimize water-resource depletion, we should shift our management objectives from locating and developing new supplies to augmenting, conserving, and reallocating existing supplies (Molle 2003). At the same time, societal objectives are evolving to value water for nontraditional uses, such as maintaining instream flows for aquatic ecosystems. Future groundwater management will have to address these multifaceted challenges. Augmenting supplies can mean improving water quality or increasing water quantity. Depletion due to quality considerations can often be overcome by treatment, whereas large volumetric depletion can only be alleviated by decreasing discharge or increasing recharge (Ali et al., 2011; Konikow and Kendy, 2005). Conserving groundwater by reducing pumpage can be accomplished through administrative, legislative, or management controls, including economic incentives to reduce demand.

In groundwater management, reallocation of water resources may be important. Effective reallocation requires rules to ensure fairness and minimize damages. When large-scale groundwater development began, no institutional mechanisms were in place to control the amount of withdrawals. In contrast to large-scale surface-water systems, which are centrally managed, groundwater supplies are mostly "managed" by individual users. Thus, groundwater development has been largely unregulated, even in many water-scarce areas.

Some regions, particularly in semi-arid and arid climates, may follow the lead of Saudi Arabia, which abandoned its goal of grain self-sufficiency through irrigated agriculture when groundwater mining could not be sustained. In other areas, large-scale water transfer projects might maintain activities and populations that depend on or benefit from the depletion of groundwater resources, even at the expense of environmental impacts in the water-exporting basin.

"Virtual" water imports and exports in the form of food-grain may be a solution in the face of groundwater depletion. For example, analyses of projected water supply and demand scenarios indicate that conventional approaches of augmenting and conserving irrigation water are insufficient to sustain agricultural water use on the North China Plain. Instead, Yang and Zehnder (2001) suggest reallocating irrigation water to urban and industrial use, retiring irrigated land, and importing grain. Ultimately, global reduction in groundwater depletion rates will likely translate to reduced crop production.

In near future, groundwater depletion will likely continue to grow, but at a reduced rate. The change in trend is already in evidence in several depleted aquifers in the western US, Asia, and other parts. The challenge for hydrogeologists is to develop and apply innovative technical approaches, built upon a solid scientific foundation, that credibly inform society of the impacts and alternatives to groundwater depletion.

In view of the exiting status of water resources and increasing demands of water for meeting the requirements of the rapidly growing population of the world as well as the problems that are likely to arise in future, a holistic, well planned long-term strategy is needed for sustainable water resources management. The water resources management practices may be based on 'increasing the water supply' and 'managing the demand under stressed condition.' Data monitoring, processing, storage, retrieval and dissemination constitute the very important aspects of the water resources management. The data are also required for the planning and design of the water resources.

In addition to this, now-a-days decision support systems are being developed for providing necessary inputs to the decision makers for water resources management. Also, knowledge sharing, people's participation, mass communication, and capacity building are essential for effective water resources management.

Relevant Journals

International Journal of Water Resources Development
Irrigation & Drainage Systems Engineering
J. of Natural Resources Dev. (JNRD)
Journal of Advances in Natural Sciences
Irrigation Journal

References

Ali, M. H. &Adham, A. K. M. (2007). Impact of climate change on crop water demand and its implication on water resources planning.*J. Agrometeorol.*, *9*(1), 20-25.

Ali, M. H., Abustan, I., Rahman, M. A. & Haque, A. A. M. (2011). Sustainability of Groundwater Resources in the North-Eastern Region of Bangladesh. *Water Resour. Manage.*,*26*, 623–641.

Kendy, E., Molden, D. J., Steenhuis, T. S. & Liu, C. (2003). Policies drain the North China Plain: Agricultural policy and groundwater depletion in Luancheng County, 1949–2000. *Int Water Manage Inst Res Rep*, *71*.

Konikow, L. F. &Kendy, E. (2005). Groundwater depletion: A global problem. *Hydrogeol J.*, *13*, 317–320.

Molle, F. (2003). Development trajectories of river basins: a conceptual framework. *Int Water Manage Inst Res Rep*, *72*.

Yang, H. &Zehnder, A. (2001). China's regional water scarcity and implications for grain supply and trade. *Environ Plan*, *33*, 79–95.

Chapter 2

ASSESSMENT OF WATER RESOURCES

ABSTRACT

Together with energy, water is one of the major fuels of economic development. Water is distributed in the earth unevenly in time and space. Although the natural water is adequate in absolute terms, it is not available at the right time, place, quantity and quality. Assessment of water resources development potential is a pre-requisite for the proper planning, management and sustainable use of the resource.

Water availability has to be evaluated as to quantity, quality and costs for various sources of water. While surface water surveys are relatively easy and inexpensive, groundwater surveys are more complex and costly as they involve test drilling, pumping, and geophysical surveys, in addition to continuous data collection through yield and water level measurements on wells.

2.1. ESTIMATION OF SURFACE WATER RESOURCE

Estimation of spatial and as well as temporal distribution of water should be accounted for. For planning purpose, it should not be averaged.

First, the effective volume of water in different available water bodies (such as ponds, lakes, rivers, etc.) should be determined during different periods (dry and wet) of the year. Then, the assessment of environmental flow requirement (such as navigation, fish culture, and other purposes) should be done. Effective volume of water for a particular season is the difference between the measured volume and the environmental flow (EF) requirement, i.e.,

Effective volume of water = measured volume – EF requirement

For estimation of existing volume of water in water bodies, different available techniques (such as trapezoidal rule, Simpson's rule, etc.) may be followed. For flowing rivers and streams, stage –discharge relationship may be used.

The effective available water is the water which can be used for different purposes, such as irrigation, hydroelectric power, etc.

2.2. Investigation of Groundwater

In many developed and developing countries, there is a heavy reliance on groundwater as primary drinking supply and/or agricultural and industrial use. A rapidly growing population has increased the demand for water in both urban and rural areas. The reliance on groundwater is such that it is necessary to ensure that there are significant quantities of water and that the water is of a high quality. This urges for a systematic investigation of the groundwater resource.

Groundwater may occur in the following modes:

- Near-surface groundwater (in alluvium)
- Intermediate perched groundwater (in the vadose zone), and
- Groundwater in the regional aquifer beneath the water-table.

In a particular location, it may exist in all of the three forms or any combination of the above.

Well yield, water quality, and monitoring well locations can be optimized by detailed fracture studies (mapping the fracture patterns and depositional environments that will affect well yield, water quality, and off-site impact) using various methods (aerial photographs, geophysical methods or borehole logging) to identify "sweet spots." Proper Well location can be critical in preventing a negative impact on environment and providing long-term life of a well. Groundwater investigation may need to carry out with the following objectives (a single *or* in combination):

- To describe the hydrogeologic units (aquifers and aquitards)
- Characterization of the aquifer
- To assess groundwater availability as both a source of potable water and as source of discharge to lakes, streams, and wetlands
- To assess the magnitude, sustainability and origin of 'deep' groundwater resources
- To examine both current and future groundwater availability
- To develop a picture of the current groundwater quality
- To assess future groundwater quality as urban and rural development continues
- To quantify a water budget
- To relate existing groundwater quality conditions to land uses and development densities; and
- To project how future land uses could impact groundwater availability and quality.

2.2.1. Planning an Investigation

The plan generally consists of a detailed outline of the objectives, scope, level of detail procedures, available equipment, existing and potential problems, necessary data, well-thought-out schedule of tasks, and the resources available.

Hydrogeologic investigation is generally complex and interdisciplinary, and requires expertise from a number of different fields. In order to properly plan a hydrogeologic site investigation, the purpose of the investigation, the general geologic and hydrologic characteristics of the site, and the management constraints under which the investigation is to take place (financial and time restraints, availability of necessary equipment, availability of expertise) should be well understood by the personnel involved in the project.

2.2.2. Steps Involved in a Site Investigation

The general steps in groundwater and hydrogeologic investigations include:

- Field reconnaissance
- Literature search
- Assessment of additional data requirement
- Planning field study/conceptual model development
- Materialization of field study
- Synthesize/evaluation of field data
- Data analysis and reporting

Field Reconnaissance

The following points which should be recorded in field reconnaissance:

- Topographic features
- General character of the local geology
- Nature and location of any significant impermeable areas
- Location of any potential surface and subsurface contamination (including volume and nature)
- Location of surface water (including nature, volume and flow)
- Flow rates of existing wells (if any)

Literature Search

Existing data should be availed first to any hydrogeologic site investigation. Much of the data necessary for developing a conceptual model may already have been collected during different previous investigations. Data can be obtained from research papers, reports (from Geologic and Natural Resources Departments, Public Works Departments, Water Resources Departments), local/federal agencies, and private organizations (specially pump drilling/ boring agencies and agricultural consultancy agencies).

Data that should be reviewed include:

- Previous boring log data of tubewells (if any)
- Previous investigations of aquifer and surface water
- Regional hydrogeologic report
- Geophysical data
- Production well data
- Topographic, geologic, and hydrogeologic map
- Soil map

- Land-use map
- Long-term climatic data (specially rainfall)

Topographic maps aid in delineating drainage areas, locating desired cross-sections, and locating boundaries for other maps (including geologic, depth to water, flow gradients, recharge and discharge areas, and other related features).

Aerial photos are often used as substitutes for topographic maps. Multiple photographs may be used with a stereoscope to obtain a three-dimensional view of the area.

Geologic maps and sections are helpful when complex geologic structures and variances occurs. When accompanied by analysis reports, they aid in locating aquifers, water level conditions, structural and stratigraphic control of water movement, and other related factors.

Water-table contour maps are similar to topographic maps, the difference being that they show water-table elevations as opposed to ground elevations.

Cross-sectional maps are developed using borehole data. The vertical stratigraphy of the subsurface is mapped out using multiple boreholes spaced in a horizontally planar manner.

The hydrographs may be for individual well (changes in water level over time), river/canal flow, and depth to water-table.

Fence diagrams are maps similar to cross-sectional maps, except they illustrate the surface and subsurface in three dimensions.

Determination of Data Requirement and Selection of Method

Based on the specific object of the investigation (and general objectives, if any), data requirement from the detailed investigation should be assessed. The investigator should choose the investigation method based on the data requirement, taking into consideration that which method will provide the most valuable data within time and cost constraints.

The following data may be needed:

Aquifer materials properties- Porosity, hydraulic conductivity, transmissibility, storativiity/specific yield, isotropy, etc.

Boundary conditions- Depth to bedrock, impermeable layer boundaries, fluxes, heads, natural water bodies, withdrawal wells, infiltration trenches, etc.

Water budget – Inflows and outflows such as surface infiltration, lateral boundary flux, leakage through confining units, withdrawals and injections.

Graphic descriptions of the conceptual model can include simplified hydrogeologic cross-sections, potentiometric surface maps, structure maps of hydrogeologic units, three-dimensional graphics, and schematic water balance diagram. Graphics may complement a written description.

2.2.3. Approaches of Investigation

Different approaches and methods have been used to investigate the groundwater resource throughout the world. The approaches include:

- Mechanical approach (Drilling bore-hole)
- Geo-physical approach
    - Electrical
    - Electro-magnetic
    - Seismic
    - Gravity
    - Magnetic
    - Ground penetrating radar
    - Borehole
    - Other approaches (magnetic resonance, Induced potential, etc.)
- Aerial photograph

The choice of a method depends on the availability of the respective instruments, technical expertise, cost, practical implementation difficulties, required precision of the results, capabilities or limitations of the method, and physiographic, socio-economic and socio-environmental factors.

2.3. MECHANICALAPPROACH

Mechanical approach is the direct method of groundwater investigation. It involves drilling, constructing boreholes, sampling and logging of boreholes to depths ranging from surface to target depth. Drilling method for investigations often aimed at resolving the origin and resource potential of 'deep' groundwater. In addition, it is also useful for chemical, isotopic and age dating of groundwaters obtained from the boreholes to better define the age, extent and origin of the deep water.

A typical borehole is indicated diagrammatically on **Figure 2.1**.

Drilling of Exploratory Bore Wells

As a part of investigation program, exploratory bore holes are periodically drilled with the following objectives:

(i) To explore the sub-surface lithological characteristics
(ii) To arrive at the sub-surface hydrogeological parameters
(iii) To find out the quality of groundwater at different aquifers.

In essence, boreholes are drilled to whatever depth the experts and drillers believe is necessary (normally up to a maximum depth of 750 feet). After test drilling, the boreholes are fully lined and grouted throughout the depth to avoid cross-contamination with locally sourced water. Water samples are obtained from constructed boreholes to be drilled at locations chosen as being the most likely to yield water.

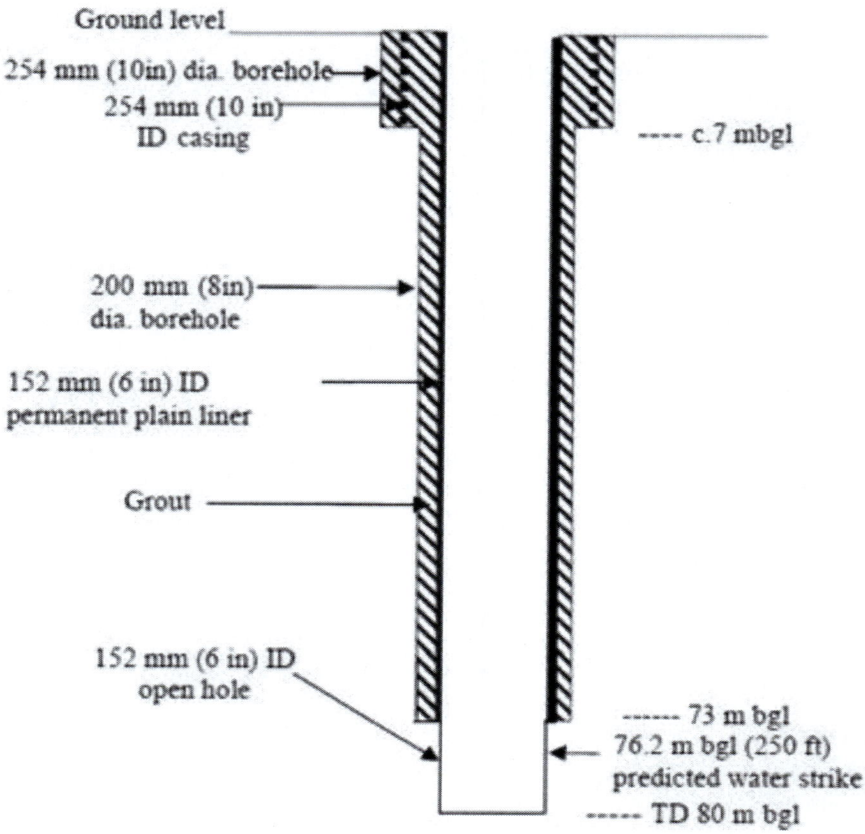

Figure 2.1. Schematic of a borehole.

Avoiding Cross Contamination during Drilling

It is necessary to specify a design for investigation boreholes that would permit water samples to be obtained from the shallow productive horizons whilst drilling, and then seal out those horizons before drilling deeper to obtain uncontaminated water samples from the greater depths. This involved drilling to a depth a few metres above that which the 'underground stream' is predicted by the water experts and well drillers, installing casing and filling the borehole with impermeable grout to provide a seal that would prevent any groundwater originating from the shallow productive fractures from entering the borehole. Once the grout had set, the borehole would be redrilled back through the casing and into the underlying rock to the required depth. Thus it would be possible to obtain groundwater samples at selected depths throughout the drilling process whilst preventing cross contamination between shallow and deep productive horizons.

Limitations

For irregular land surface and uneven/discontinuous lithological condition, it requires a greater number of boreholes for complete information; which is very costly to perform and time consuming.

Sampling Interval and Representation

Sampling interval depends on the specific objective and the pattern of lithological condition. If no abrupt change is occurred, sampling at 10 ft interval is sufficient. At transition, frequent sampling may be needed. The soil samples are analyzed in the laboratory for particle size distribution. For aquifer materials, it is analyzed for specific yield, well screen design, and envelop material design. Water sample should be collected during drilling and development for quality analysis.

2.4. GEO-PHYSICALAPPROACH

The main use of geophysics in the geosciences is for hydrocarbon exploration typically at depths greater than 1000m. Near-surface geophysics for groundwater investigations is usually restricted to depths less than 250m below the surface. Groundwater applications of near-surface geophysics include mapping the depth and thickness of aquifers, mapping aquitards or confining units, locating preferential fluid migration paths such as fractures and fault zones and mapping contamination to the groundwater such as that from saltwater intrusion. The use of geophysics for groundwater studies has been stimulated in part by a desire to reduce the risk of drilling dry holes and also a desire to offset the costs associate with poor groundwater production. It also provides useful parameters for hydrogeological modelling of both new groundwater supplies and for the evaluation of existing groundwater contamination. Geophysical methods can be used in groundwater related investigations to:

- Locate geological structures such as faults and formation contacts
- Identify gravel channels within finer grained materials
- Define watertable in unconsolidated sediments
- Identify subsurface voids
- Map higher permeable areas within aquifers
- Define the extent of weathering, fracturing and faulting
- Define the extent of protective clay layers
- Map soil conductivity
- Identify freshwater-seawater interface/sea-water intrusion
- Provide complimentary data for correlating borehole and monitoring well data

Principle of Geophysical Approach

For soils and rocks, the properties can be divided into a framework or matrix component and the pore content component. All geophysical techniques measure variations in a material's physical properties. Different materials exhibit different parameter signatures such as their resistivity or its inverse, conductivity, acoustic velocity, magnetic permeability and density. These parameters are influenced by the mineral type, grain packing arrangement, porosity, permeability, and pore content (i.e., gas or fluid type). In general, no one property is unique to any materials; rather a material is described by ranges of each property. In most geophysical surveys therefore it is important that the changes or contrasts in geophysical parameters are measured and that the target shows large property differences with surrounding material.

2.5. ELECTRICALMETHOD

Electrical techniques have been extensively used in groundwater geophysical investigations because of the correlation that often exist between electrical properties, geologic formations and their fluid content. This method is also termed as electrical conductivity or resistivity method, and resistivity imaging method.

Principle of the Method

Most electrical techniques induce an electrical current in the ground by directly coupling with the ground. Electrical conductivity, or its inverse resistivity, is the proportionality factor relating the electrical current that flows in a medium to the applied electric field. The resulting electrical potential is then used to measure the variation in ground conductivity, or its inverse, resistivity. Different materials, and the fluids within them, will show different abilities to conduct an electric current.

This method has also been applied to assess water flow in the non-saturated zone of the soil. The resistivity decreases as moisture content or the dissolved solids in the interstitial water increases.

2.6. ELECTROMAGNETIC METHOD

While the final output of electromagnetic method is similar to that of electrical techniques, several advantages with the electromagnetic techniques result in an increased resolution and more cost-effective application.

Two types of electromagnetic survey are currently practiced:

(i) Time domain electromagnetic (TDEM) surveys which are mainly used for depth soundings and recently in some metal-detector type instruments, and
(ii) Frequency domain electromagnetic (FDEM) surveys that are used predominantly for mapping lateral changes in conductivity.

In both electromagnetic survey techniques no direct contact is made by electrodes with the ground and thus the rate of surveying can be far greater than for electrical techniques where electrode probes must be placed in the ground for every measurement.

Principle of the Method

Both techniques measure the conductivity of the ground by inducing an electric field through the use of time varying electrical currents in transmitter coils located above the surface of the ground. These time-varying currents create magnetic fields that propagate in the earth and cause secondary electrical current to flow in the subsurface (the process is known as electromagnetic induction). The secondary current can be measured either while the primary field is transmitting (during frequency domain surveys) or after the primary field has been switched off (for time domain surveys). The receiving coil responds to an electromotive force generated by the resultant of the primary and secondary fields. Buried metal objects (e.g., drums, pipelines, etc.) and water bearing strata produce characteristic anomalies which result from the geometrical relation between the object and the instrument coils along with the very high conductivity of the metal.

2.7. Estimation of Groundwater Potential

Groundwater assessment requires conceptualization of the physical process and methodology for assessment. The following approaches may be employed to estimate the groundwater potentials:

2.7.1. Quantitative Estimation of Groundwater

Assessment of extractable groundwater quantity is the main objective of groundwater investigation. From the areal and depth information, quantification of groundwater is possible knowing the *specific yield* of the formation. *Specific yield* of the aquifer is the volume of water per unit volume of aquifer that can be extracted by pumping (or gravity drainage). Specific yield and porosity are related as follows:

$$\text{Porosity} = \text{specific yield} + \text{specific retention}$$

Specific yield (S_y) is clearly an important factor in water availability, and is the factor that is used to convert saturated thickness (T_s) to the actual volume of groundwater available. Extractable water can be calculated as follows:

$$V = A \times T_s \times S_y \qquad (2.1)$$

Where,
V = total volume of extractable water (m^3)
A = saturated area of the formation (arial extent) (m^2)
T_s = saturated thickness of the formation/aquifer (m)
S_y = Specific yield of the formation (m^3/m^3)

For a stratified formation, or total yield from a well field, the total amount can be computed as:

$$V = \sum_{i=1}^{n} (A_i \times T_{S_i} \times S_{y_i}) \tag{2.2}$$

Where, *i* is the number of strata *or* aquifer.

Specific yield is generally expressed as the percentage of the total volume (volume of saturated formation) (%), or volume per unit volume (m^3/m^3). For alluvial deposits (medium to coarse sand, sandstone, sand and gravel), the specific yield varies from 20 to 35%. For fine to medium sand, it varies from 10 to 20%.

2.7.2. Groundwater Availability for Pumping in Terms of Potential Recharge

Groundwater availability for pumping in terms of potential recharge may be estimated from simplified hydrological balance equation:

$$P = PET + U + R_p$$

Or,

$$R_p = P - PET - U \tag{2.3}$$

Where,
P = Rainfall
PET = potential evapo-transpiration
U = Runoff
R_p = Potential recharge

From the above equation, potential recharge is the excess of rainfall over runoff and potential evapo-transpiration. The estimate can be carried out weekly, monthly, seasonal, or yearly basis. Monthly basis is more judicious than the seasonal or yearly estimates.

In most cases, the surface runoff is scarce, and difficult to obtain. In the absence of data on surface runoff, runoff may be presumed as a percentage of rainfall; depending on the rainfall amount, intensity of rainfall, topography, physical and hydraulic characteristics of the topsoil and geo-hydrologic condition (normally, 20 ~ 50%). The runoff rate depends on antecedent soil moisture, land use, surface cover, slope and topographic condition. A complex model, which is capable of incorporating interacting factors on the amount of runoff, may assess the runoff more closely.

Recharge can also be estimated using other direct and indirect methods (see Chapter 3).

2.7.3. Groundwater Availability in Terms of Safe Yield

Groundwater availability in a catchment/basin or region may be estimated based on safe yield concept. According to this concept,

Safe yield = maximum allowable fluctuation of water-table (MAWT)
× specific yield of the aquifer

Here, MAWT refers the difference between maximum allowable depletion or lowering of water-table (WT) during pumping (considering environment, other ecosystem), and the present depth to water-table; i.e.,

MAWT = Maximum allowable depletion of WT – present depth to WT

The use of historically observed maximum fluctuation in annual water-table (WT) in safe-yield or recharge estimation may result in an underestimated value of recharge, if there is rejected recharge. The annual fluctuation of WT can be increased by withdrawal of groundwater for irrigation during dry season, thus creating further scope for increased recharge (i.e., the potential recharge). The potential recharge may be defined as the mean annual/seasonal volume of surface water that could reach the aquifer, which is the sum of actual recharge to the aquifer and the rejected recharge (if any). Rejected recharge is that fraction of water available at the surface which can not percolate because the water-table is near/at the surface.

Determination of Specific Yield

Specific yield can be determined by free drainage from the collected aquifer materials (through bore logs). It is an important parameter in 'safe yield' estimation or 'groundwater potential' estimation using this approach. It depends on the relative percentage of ingredients. The value of specific yield varies widely within the physiographic zones and even in a zone. For accurate estimation of groundwater potential, the specific yield should be determined for each zone and the mean of representative number of samples should be used, so that the data represent the actual aquifer characteristics.

For each of the above approaches, the total area of a basin or a region should be divided into suitable physiographic units on the basis of some criteria such as geological condition, topography, physical and hydraulic characteristics of the topsoil, and/or depth of flooding. Available and potential recharge for each unit should be calculated and weighted average should be determined.

2.7.4. Water Budget Approach

The water available for sustainable use must be balanced against the volume of water input to the system. Therefore, the water available for sustainable use is independent of the recharge volume, but is dependent on the response of the aquifer system to the volume of water withdrawn.

Expression of Water Budget

The components of the water budget (input, output and changes in storage) can be expressed mathematically. If the system is in equilibrium, the water budget can be expressed in its simplest form by the mass balance equation:

$$O = I \pm \Delta S \quad (2.4)$$

where

O= water output or discharge;
I = the water input or precipitation; and
S= water storage,
ΔS = change in storage

I—Input

Rainfall makes up 100% of the water input in some aquifer system, where no major river systems traverse or otherwise deliver water to the catchment. Input may be formed from canal or river system, where available. Groundwater recharge from the adjacent aquifers can be considered negligible, where the groundwater systems beneath the area are elevated above the surrounding systems; otherwise may be significant.

Estimation of Potential Recharge

Details of groundwater recharge have been discussed in *Chapter-3*. In order to know the potential recharge in a region, inter-relationship between rainfall, runoff and infiltration, percolation, ET, and groundwater level response to rainfall should be studied. Potential recharge is not only dependent on surface water balance component and recharge opportunity, but also dependent on the withdrawal from the aquifer. Potential recharge can be mathematically expressed as:

$$R_p = f(S_s, S_b, P, R, ET, D, \ldots) \quad (2.5)$$

Where,
R_p = potential recharge
f = function
S_s = soil surface condition
S_b = subsurface soil condition
P = rainfall characteristics (intensity, duration)
R = surface runoff
ET =evapotranspiration
D = discharge or withdrawal from the aquifer

Investigations should be carried out at micro level and basic data like infiltration through rainfall, seepage from applied water for crops, from tanks, from canals etc. These parameters will help in the assessment of groundwater potential, and the block-wise groundwater assessment can be made at macro level for the entire catchment.

O—Output

The output may be comprised of direct surface runoff or stormflow, groundwater discharge, evapotranspiration, and groundwater consumption.

Surface Runoff

Surface runoff is the volume of water that runs directly off the land surface into streams without infiltrating into the groundwater system. The surface runoff leaving a catchment can be estimated by subtracting stream baseflow from the mean annual streamflow.

Groundwater Discharge

Groundwater discharge can be assumed to be equal to the groundwater recharge (if there is no pumping), which is assumed to be equal to the stream base-flow.

Evapotranspiration

Evapotranspiration (ET) is the combined effect of evaporation from wet surfaces and transpiration from plants. The ET can be estimated using various available methods and tools.

S - Storage

Changes in aquifer storage are reflected in water-table elevation changes for unconfined aquifers and the piezometric surface elevation for confined aquifers. Changes in aquifer storage are also reflected in changes in spring production and stream flow. Changes in spring production and stream flow are very difficult to measure. Changes in water-table elevation, on the other hand, can be determined if there are sufficient data available.

Water level measurements can be made on wells on the catchment. Therefore, while changes in groundwater storage cannot be estimated in absolute terms, it is possible to determine if there are any declines that would indicate a loss of storage, or increases that would indicate an increase in storage, by examining changes in water-table elevation. Statistical significance test of water-table elevation change (for long-term trend) can be performed.

Estimation of Absolute Change (Amount)

Change in absolute volume in an aquifer system can be estimated as:

$$\Delta S_v = \Delta S_h \times S_y \times A \tag{2.6}$$

Where,

ΔS_v = change in water volume (m^3)

ΔS_h = change in water level in observation well (m)

S_y = specific yield of the aquifer (%)

A = area of the aquifer (m^2)

2.8. Groundwater Development Potential and Issues in Saline/Coastal Areas

The development of freshwater in coastal saline area requires careful observation. Freshwater in the saline and coastal area may occur in single or two zones (shallow and deep freshwater) (Saleh and Nishat, 1989; Rashid, 2008). Localized pressure reduction as a result of tube-well abstraction may cause upconing of saline water interface. To mitigate sea-water intrusion in aquifer (salinization of the aquifer), it is essential to control groundwater abstraction, have to adopt suitable well design procedures and should have a thorough knowledge of the hydrogeology of the aquifer.

2.9. Environmental Flow Assessment

Environmental flow is the water regime that should be provided within a river, wetland or coastal zone to maintain ecosystems and their benefits. It should be ensured to maintain the conditions that supply goods and services and to ensure biodiversity.

The following methods are commonly used to assess environmental flow:

(i) Building block method
(ii) Drift method

Building Block Method (BBM)

In this method, a certain ecological status of a watercourse or river or wet-land is maintained. The desired/targated condition of the river is identified. Environmental flow regime is then constructed based on premise that riverine species rely on basic elements of flow regime (building block).

The principal phases of building block method are:

(a) Information gathering on watercourse or river system, normally undertaken by a team of experts.
(b) Discussion and exchange opinion among agency representative, water managers, engineers and scientific experts to identify flow regimes. Optimising the environmental flow consideration with engineering considerations

Information gathering involves:

- delineate the study area
- recording the present condition of river for overall riverine habitat
- determination of importance of study area at local, regional and international scale.
- assessing the quality of water
- biological surveys (may be at selected points)
- identification of building block method reaches and sites
- documentation of natural and present daily flow regime
- analysis of stage-discharge curve

The objective of the exchange of opinions among different experts is to identify and conclude on the recommended flows for the river. The experts should visit the site. Attention should be given to the flow features that are considered most important for maintaining or achieving the desired state of the river. These flow features are the building blocks which are constructed to create the in-stream flow requirements. Required flows are identified month-wise, considering low and high flows.

In final stage, recommended flow regimes are integrated with engineering concerns for the river. Two or three possible flow states with probable economic and social consequences are developed.

The building block method is costly for data collection and employment of experts. The final selection is dependent on professional judgment and experience.

Drift Method

It is a scenario based method. Scenario is used to identify tradeoffs that can be used in negotiations between water users to balance the needs of the ecosystem with other needs. A set of scenarios (produced as output from different water levels) or options for a river is used in decision and policy making.

Relevant Journals

International Journal of Water Resources Development
Water Resources Research
Water Research
Water Resources Management
Journal of American Water Resources Association
J. of Natural Resources Dev. (JNRD)

EXERCISES

1. Briefly narrate the principle of quantifying effective/usable amount of surface water
2. Narrate the issues in planning a groundwater investigation
3. What are the steps in a site investigation?
4. Discuss the different approaches of groundwater investigation
5. Briefly discuss different approaches for determination of groundwater potential
6. Discuss the issues of groundwater development potential in coastal aquifers
7. What is Environmental flow? Narrate the procedure for its assessment.

Chapter 3

ESTIMATION OF GROUNDWATER RECHARGE

ABSTRACT

To properly manage groundwater resources, managers need accurate information about the inputs (i.e., recharge) and outputs (i.e., pumpage and natural discharge) within each groundwater basin, so that the long-term behavior of the aquifer and its sustainable yield can be estimated or re-assessed. Reliable recharge data may be used as the basis for studying potential climate change impacts on groundwater resources. Recharge is a major component of the groundwater system, and has important implications for shallow groundwater quality. The variability of recharge is the key to understand the susceptibility of aquifers to contamination from surface-derived sources. In this chapter, details of groundwater recharge – from basic concept to detail measurement procedures, limitations, selection process of a method, challenges in generalized recharge model, and workout problems on different aspects of recharge estimation are described.

3.1. CONCEPT, PURPOSE AND SIGNIFICANCE OF RECHARGE ESTIMATION

Concept

Generally speaking, the part of infiltrated water that moves downward and reaches to aquifer *or* water-table is termed as recharge. The horizontal movement of water from depressions, lakes, rivers, etc. to water-table is also part of total recharge. That is, recharge can be broadly defined as water that reaches an aquifer from any direction (up, down, or laterally). The part of infiltrated water that does not reach to aquifer *or* water-table, contributes to soil moisture storage.

The amount of water that may be extracted from an aquifer without causing depletion is primarily dependent upon the groundwater recharge. Rainfall is the principal source for replenishment of moisture in the soil water system and recharge of groundwater. Other sources include recharge from rivers, streams, irrigation water, etc. Moisture movement in the unsaturated zone is controlled by suction pressure, moisture content and hydraulic conductivity relationships. The amount of moisture that will eventually reach the water table is defined as natural groundwater recharge, which depends on the rate and duration of rainfall, the subsequent conditions at the upper boundary, the antecedent soil moisture conditions, the water table depth and the soil type.

The recharge may be natural (as that of the mentioned cases) and human induced (artificial). Human induced recharges include construction of ridges around the field plot, digging canals, water storages, and construction of recharge well (opposite in nature of discharge well), etc. to facilitate more water to move downwards.

Purpose

Typical purposes of recharge estimation include:

(i) Evaluation of water balance at catchment scale
(ii) Estimation of sustainable aquifer yield
(iii) Determining safe limit of groundwater withdrawal
(iv) Determining possibility of groundwater contamination (from point *or* distributed sources)

Significance

Quantification of the rate of natural groundwater recharge is a pre-requisite for efficient groundwater resource management. It is particularly important in regions with large demands for groundwater supplies, where such resources are the keys to economic development.

Estimation of recharge is critical in any analysis of groundwater systems and the impacts of withdrawing native water from them. In water-resource investigations, groundwater models are often used to simulate the flow of water in aquifers, and, when calibrated, may be used to predict long-term behavior of an aquifer under various management schemes. Without a good estimate of recharge and its spatio-temporal distribution, these models become unreliable. Accurate estimates of recharge and recharge mechanisms are also necessary to assess the risk of groundwater contamination, particularly diffuse agricultural contamination (e.g., from nitrates and pesticides) and concentrated *or* point-source of contamination (e.g., landfill). Clearly, understanding recharge is critical to managing most groundwater systems.

3.2. Relevant Terminologies

In many unsaturated-zone studies, terms such as net infiltration, drainage, or percolation are used to describe water movement below the root zone, and these are often equated to recharge. But only for limiting conditions, they may be equal to recharge, not for all cases. The terminologies relevant to recharge are defined below.

Infiltration

Infiltration refers to the entry of water from soil surface to soil body.

Percolation

Percolation refers to the vertical movement of water through soil layers.

Seepage

Horizontal movement of water through soil layer(s) is termed as seepage.

Actual Recharge

Recharge in an area under existing conditions of the eco-system (factors which influence recharge) [i.e., under normal condition, some factors may be in limiting conditions].

Potential Recharge

The maximum possible recharge that can occur in a geological formation under non-limiting conditions of other factors (such as availability of water, infiltration barrier, aquifer fullness, etc.) is termed as potential recharge.

Direct/Diffuse Recharge

Direct or diffuse recharge refers to recharge derived from precipitation or irrigation that occurs fairly uniformly over large areas.

Indirect Recharge

Recharge that results from runoff and localization in joints, as ponding in low-lying areas and lakes, or through the beds of surface-water courses.

Localized/Focused Recharge

Localized recharge refers to concentrated recharge from small water-bodies such as ponds, depressions, joints *or* rivulets.

Artificial Recharge

Deliberate act of adding water to an aquifer by means of a recharge project, also the water so added. Artificial recharge can be accomplished via injection wells, spreading basins, or in-stream projects.

Natural Recharge

Naturally occurring water added to an aquifer. Natural recharge generally results from snowmelt and precipitation or storm runoff.

Induced Recharge

Recharge to groundwater as a result of lowering the groundwater level below the surface-water level.

Base-Flow

Stream-flow derived mainly from groundwater seepage into the stream.

Darcy's Equation or Law

A formula stating that the flow rate of water through a porous medium is proportional to the hydraulic gradient. The factor of proportionality is the *hydraulic conductivity*.

Deep Drainage

Drainage of water below the root zone.

Drainage Basin

The surface area which drains to a specified point on a watercourse, usually a confluence of streams or rivers. It is a hydrologic unit (consisting of a part of the surface of the earth) covered by a drainage system consisting of a surface stream or body of impounded surface water plus all tributaries.

Groundwater Basin

A geologically and hydrologically defined area which contains one or more aquifers.

Hydrologic Budget or Water Budget

An accounting of the inflow to, outflow from, and storage in a hydrologic unit such as a drainage basin, aquifer, soil zone, lake, or reservoir.

Piston Flow or Plug Flow

Purely advective flow without dispersion or diffusion of the dissolved components.

Preferential Recharge

Recharge that takes place preferentially through macropores, as opposed to *diffuse recharge*, which takes place through the entire vadose porous medium.

Recharge Area

The area that contributes water to an aquifer. It is normally considered to be the natural area of recharge, as contrasted with a constructed recharge basin.

Rejected Recharge

Potential recharge that exceeds the rate of flow through an aquifer that is already overfull, and as a result is rejected.

Residence Time

The length of time between the input of water as infiltration or recharge and its output as runoff or discharge. Also known as *transit time* or *turnover time*.

Residual

In the case of recharge, the remainder of all other hydrologic components in the *water-balance* equation.

3.3. Sources and Mechanism of Recharge

Continued availability of groundwater (for irrigation or other users) is dependent on annual recharge of the aquifers. While estimating natural groundwater recharge, it is essential to have a good idea of the different recharge mechanisms and their importance in the study area.

The location of the recharge zone is generally controlled by the outcropping of a more permeable unit at the surface. However, the recharge may occur uniformly across the recharge zone or there may have preferred pathways for infiltration recharge. The recharge zone mainly comprises hilly terrain underlain by soils and regolith that has very slow permeability, and there may be a slow rate of recharge through these soils. Alternatively losses of water from stream-flow in the valley bottoms, or from soils in lower slopes, may be a preferred pathway since these areas tend to be underlain by more permeable soils than the hilly terrain, and by more freely draining gravels.

The sources of recharge to a groundwater system may include both natural and human-induced phenomena. Natural sources include recharge from precipitation, lakes, ponds and rivers (including perennial, seasonal, and ephemeral flows), and from other aquifers. Human-induced sources of recharge include irrigation losses (both from canals and irrigated fields), leaking water mains, sewers, septic tanks, and over-irrigation of parks, gardens, and other public amenities. Recharge from these sources (from percolation of precipitation) is termed as *direct recharge,* and *recharge* from runoff ponding *is termed as indirect recharge.*

Direct mode of recharge is spatially distributed (diffuse) and results from widespread percolation through the entire vadose zone. It is typical of humid climates because frequent, regular precipitation maintains a high water content in the soil, so that there is little additional storage capacity in the vadose zone; thus, infiltration can be routed quickly through the vadose zone to the saturated zone. This recharge raises the water table, which leads to increased stream-flow. Thus, in humid climates, flowing perennial streams are typically groundwater discharge areas sustained by diffuse recharge in the basin.

Indirect recharge results from percolation to the water-table following runoff and localization in joints, as ponding in low-lying areas and lakes, or through the beds of surface-water courses. In arid and semi-arid regions, localized and indirect recharge are often the most important sources of natural recharge.

Percolation to the water table from stream-beds may takes two forms - saturated flow and unsaturated flow, depending on whether there is a saturated connection between the stream and the water-table. In mountainous areas, infiltration from mountain streams may be of complex processes of unsaturated and saturated flow in fractured rocks, as well as infiltration along channels flowing across alluvial fans.

The recharge which takes place through the entire vadose porous medium is termed as *diffuse recharge*. *Macropore flow* occurs in soils having pores with a diameter or width larger than 3 mm, such as cracks in clay soils, rock fractures, fissures in sediments, solution channels, worm holes, and old root channels. In macropore flow, the effects of capillarity are no longer felt, and the flow process is dominated by viscous forces and gravity. Flow through macropores also is known as *preferential* or *bypass flow*, and the resulting recharge is called *preferential recharge*, which preferentially takes place through such macropores, as opposed to diffuse recharge. The velocity with which water moves through macropores from the soil surface to the water table often is several orders of magnitude higher than through the soil matrix. Saturated flow through macropores can be quantified using *Poiseuille's equation* as opposed to Darcy's equation for diffuse flow.

3.4. Factors Affecting Recharge

The climate, hydrology, geomorphology, and geology of a region control the recharge rate and total recharge. Information on topography, land-use/land-cover, and soil types is combined to define geomorphic systems. Recharge can vary substantially both within and between watersheds because of variations in precipitation, geological and geomorphologic settings. In arid climates, long-time series are needed to assess mean annual recharge rate.

In general, the following factors influence the recharge rate:

(i) Soil factor
(ii) Topography
(iii) Land-surface/vegetation
(iv) Sub-surface geology
(v) Climate
- Rainfall amount and its distribution
- Evaporative demand of the atmosphere

(vi) Depth to aquifer
(vii) Storage capacity of the aquifer
(viii) Existence of water bodies

Soil Factor

The factors which influence the infiltration rate, also influence the recharge rate. These include soil type (textural class), bulk density, organic matter content etc. Soil texture and permeability are important because coarse-grained soils generally result in higher recharge rates than do fine-grained soils. Cook et al. (1992) noted an apparent negative correlation between clay content in the upper 2 m and the recharge rate.

Topography

Recharge is generally considered to occur in topographic highs and discharge in topographic lows in humid regions, whereas in arid alluvial-valley regions recharge is usually focused in topographic lows, such as channels of ephemeral streams.

Delineation of hydrogeomorphic settings on the basis of topographic attributes – including slope classes, breaks in slope, curvature, and elevation – is greatly facilitated by the use of geographic information systems (GIS) and digital elevation models and is used widely in many countries (like Australia, USA, Malaysia).

Land-Surface and Vegetation

For rainfall to percolate into the aquifers, any soil moisture deficit needs to be overcome. Then surplus water is available to runoff or recharge groundwater. Changes from short to tall vegetation reduce the total surface water yield and this same effect is likely to apply to the amount of water available for groundwater recharge.

If the surface slope of the land is very steep, water moves quickly to downward, thus there is less time or opportunity to enter into the soil. The reverse is true for flat land.

The presence and type of vegetation is of paramount important to the physical processes occurring at land surface. In particular the runoff, heat and water vapor are controlled by soil cover and leaf area index. The surface energy and water balance varies with vegetation dynamics. Vegetation cover is important in assessing recharge potential at a site. Recharge is generally much greater in non-vegetated than in vegetated regions and greater in areas of annual crops and grasses than in areas of trees and shrubs.

The impact of vegetation was demonstrated in Australia, where replacement of deep-rooted native Eucalyptus trees with shallow-rooted crops resulted in recharge increases of about two orders of magnitude (<0.1 mm/year for native mallee vegetation to 5–30 mm/year for crop/pasture rotations) (Allison et al. 1990).

Sub-Surface Geology

Recharge rate is largely influenced (and limited) by the ability of the subsurface strata (up to the aquifer) to transmit water, which is strongly related to subsurface geology. Recharge in the soil-covered areas is several orders of magnitude smaller than in the hardrock terrain.

Climate

Climate plays a major role in controlling recharge. Recharge rates are limited in large part by the availability of water at the land surface, which is controlled by climatic factors, such as precipitation and evapotranspiration, and by surface geomorphic features.

Rainfall Amount and Its Distribution

The total rainfall amount and its distribution greatly affect the recharge. Separate high rainfall events mainly contribute recharge if annual precipitation is extremely low. If other conditions remain constant, the more water is available at the soil surface, the more will be recharge. For the same amount of rainfall, recharge is affected by the temporal distribution of rainfall. If large amount of rainfall occurs within short period of time, the water excess of infiltration capacity will move as surface runoff to the river and finally to ocean. If the same rainfall amount occurs at larger time period (i.e., having low rainfall intensity) and distributed nearly regular interval, the opportunity time for infiltration will be greater and hence more recharge.

Evaporative Demand of the Atmosphere

The evaporative demand of the atmosphere controls the recharge in the way that, if the evaporation or evapo-transpiration rate is high, the surface water ceases quickly facilitating less time for infiltration or recharge. The opposite is true for the location having less evaporative demand.

Existence of Water Bodies or Streams

Recharge rates are largely limited by the availability of water at the soil surface. If the sources of recharge, i.e., surface water bodies (such as rivers, lakes, streams, low-lands, depressions) exist and the area is significant/large, a higher amount of recharge will take place. Similarly, recharge will take place from irrigated fields. In addition, irrigation return flow often contributes significant amounts of recharge.

Storage Capacity of the Aquifer

If the aquifer is characterized by smaller storage capacity and shallow water-table, it is often full and water is usually discharged through evapo-transpiration and base-flow to streams.

Depth to Aquifer

If the depth to aquifer is too high, there is little chance of recharge to take place. There are huge possibilities to divert water in other ways – such as evaporation, transpiration, base-flow to stream etc.

3.5. GW Recharge-Discharge/Withdrawal Relationship and Sustainability Issues

Under natural conditions (and before any development), groundwater recharge is balanced by groundwater discharge, i.e., *Recharge = Discharge*. As the groundwater is nearly always moving, it will naturally flow from the recharge areas to the discharge areas. The discharge from the aquifers may occur in a variety of ways such as flow to streams, lakes, and springs; water use (transpiration) by phreatophytic vegetation that draws its water from the water table or its capillary fringe; evaporation from playas and areas of very shallow water table; leakage to adjacent aquifers; or flow to the sea.

Pumping groundwater constitutes an additional withdrawal from the system. In order for the system to reach a new equilibrium (a state of *sustainability*), the pumping must either cause the recharge to increase, and/or it must cause the discharge to decrease. Groundwater pumping usually has little impact on the recharge, especially under arid and semiarid conditions with deep water-tables, because recharge is determined mostly by climatic conditions, although in areas of intense irrigation, return flows to the underlying aquifer could be significant. Pumping, however, can decrease groundwater discharge by lowering shallow water-tables, thus reducing groundwater evapo-transpiration and seepage to streams, springs, lakes, or wetlands.

In order for a groundwater system to be *sustainable*, pumping must be balanced by an equal capture of discharge and/or recharge. If pumping exceeds the total amount of natural recharge or discharge to the system, groundwater *mining* occurs, and the system is no longer sustainable. However, even without regard to the environment, it is not always possible to extract all of the natural aquifer recharge or discharge. In some cases, wells will run dry before natural groundwater discharges are reduced to zero. The fraction of recharge that can theoretically be extracted from an aquifer under steady-state conditions will depend on the geometry of the aquifer system, and, in particular, on the location of the pumping wells relative to the natural recharge and discharge zones. Therefore, the sustainable yield of aquifers, and thus the environmental impact of groundwater extraction, depends not only upon the volume extracted, but also on the location of pumping wells relative to recharge and discharge areas, and sometimes also on the timing of the extraction.

Prediction of environmental impacts of groundwater extraction always requires detailed investigation of natural groundwater recharge and discharge processes. It is important to note that all levels of groundwater extraction will, in the long run, result in declines of natural discharges, with consequent environmental impacts. Sometimes such impacts will be small and not readily identifiable, while in other cases, they may be much more dramatic, such as in the drying up of springs and streams However, there will always be a *time lag* between groundwater extraction and reduction in natural discharge, and therefore the current apparent health of an exploited aquifer and the ecosystems that depend upon it does not necessarily indicate that the situation will be sustainable in the longer term. The task of groundwater managers is to determine what limits of environmental impact are acceptable to the community and to manage extraction to maintain impacts within those limits.

3.6. Functional Form of Recharge and Limiting Conditions

3.6.1. Functional Form

Recharge rate (R) in an area can be better described as:

$$R = f(S, T, V, P, C_A, G_s, D_{WT}, ET..) \tag{3.1}$$

Where,
f = Function
S = Soil factor
T = Topographical factor
V = Vegetation factor
P = Precipitation *or* Rainfall factor
C_A = storage capacity of the aquifer

G_s = Subsurface geological factor
D_{WT} = Depth to water-table
ET = Climatic Evapotranspiration (ET) demand

3.6.2. Limiting Conditions

Limiting conditions of recharge may be as follows:

(i) If T, V, G_s, C_A and D_{WT} are not limiting in the above equation (eqn.3.1), the potential recharge is determined by S (soil factor), ET and P (available water *or* rainfall at the surface) as follows:

(a) If $I_p > (P - ET)$, $R = (P - ET) = R_a$; until $(P - ET) \leq I_p$

Where,
I_p = Infiltration capacity of the soil
R_a = Actual recharge rate
P = Rainfall rate
ET = Evapo-transpiration rate

That is, when there is no ponding water on the soil surface, recharge rate, R, will be equal to the difference between water input rate (P) and evapotranspiration rate (ET), until the difference i.e., (P-ET) is less than or equal to the infiltration capacity, I_p. Here, actual recharge (R_a) is less than the potential recharge (R_p), that is $R_a < R_p$.

(b) If $I_p < (P - ET)$, $R = I_p = R_p$

Where, R_p is the potential recharge rate.
Under this condition, potential recharge occurs.

(ii) If S, T, V, C_A and D_{WT} are not limiting in equation (3.1), the potential recharge is determined by G_s (sub-surface geology), ET and P (available water *or* rainfall at the surface) as follows:

(a) If $K_v < (P - ET)$, $R = K_v = R_p$ subject to $I_p > K_v$

Where, K_v is the minimum vertical hydraulic conductivity of any layer above the water-table (or aquifer) (which is the characteristic of sub-surface geology). That is, in this case, the recharge will be controlled by the conductivity of the less permeable layer, and it is the potential recharge of this area (i.e., $R = K_v = R_p$).

(b) If $K_v > (P - ET)$, $R = (P - ET) = R_a$ subject to $I_p > K_v$

That is, under this condition, potential recharge does not occur.

3.7. Recharge Estimation - Available Approaches and Methods

Initiating Recharge Study/Preliminary Recharge Estimate

The first step of a recharge study in an area (specially for areas where recharge has not previously been studied) should involve collecting existing data on potential controls on recharge, such as climate, hydrology, geomorphology, and geology. These data are used to develop a conceptual model of recharge in the system. The conceptual model describes location, timing, and likely mechanisms of recharge and provides initial estimates of recharge rates.

Preliminary recharge rates for a site can be estimated using available meteorological data and soil hydraulic-parameter data in unsaturated-zone models. Available hydrologic data should also be evaluated, including stream-flow data (for evaluating gaining and losing sections of streams) and water-table depth (for determining unsaturated-zone thickness).

Recharge Estimation Techniques

The recharge estimation techniques can be categorized in several ways:

A. Based on ‘ Mode of measurement’
B. Based on ‘Techniques of estimation/measurement’
C. Based on ‘Saturation condition of strata’
D. Based on ‘Suitability for spatial scale’

Under category ‘A,’ the recharge estimation techniques can be broadly classified into:

(i) Direct methods
(ii) Indirect methods

Direct methods include: Lysimeter, seepage meter, tracer technique

Indirect methods include: Water balance (or water budget) approach, numerical modeling, empirical method, heat tracer

Under category ‘B,’ the recharge estimation techniques can be broadly classified into physical, tracer, numerical modeling, and empirical approaches. Under each approach, several methods are available:

Physical –Seepage meter, lysimeter, zero-flus plane, Darcy’s law, water-table fluctuation, baseflow discharge, water budget. Physical methods rely on direct measurements of hydrological parameters or on estimates of soil and/or aquifer physical parameters.

Tracer – Stable isotope, radioactive isotope, historical isotope, environmental tracer, dye, etc.

Numerical model – Single object optimization model, multiple objects optimization model, composite model study.

Empirical method– Rely on empirical equation.

Under category ‘C,’ the recharge estimation techniques can be broadly classified into:

1. Saturated zone approach
2. Unsaturated zone approach

Saturated zone approaches – Seepage meter, surface ponding method, lysimeter method

Unsaturated zone approaches – Darcy’s law, zero-flux plane method, lysimeter, tracer techniques,

In unsaturated zone approach, estimates of potential recharge are based on drainage rate below the root zone. These techniques are mostly applied in semi-arid and arid regions, where the unsaturated zone is generally thick.

Under category 'D,' recharge estimation techniques may be classified as:

1. Local scale methods – Lysimeter method, Darcy's law approach, seepage meter
2. Basin-scale and regional methods – Water balance, watershed model, numerical simulation models, etc.

3.7.1. Water Budget (or Water Balance) Method

Water-budget methods are those that are based, in one form or another, on a water-budget equation.

Principle of the Method

In this approach, recharge is estimated as "residual" of water-budget equation, whereby all of the variables in the equation except "Recharge (R)" are measured or estimated, and R is set equal to the residual.

Boundaries Require

Water budget method requires two types of boundaries: (i) physical *or* spatial boundary, and (ii) temporal (time) boundary.

Mathematical Formulation of Water Budget Equation

Water budget of a hydrological unit (or basin) is an account of all quantities of water added to, subtracted from, and stored (within a given volume of soil) during a given period of time. The water balance is merely a detailed statement of the law of conservation of matter, which states simply that matter can neither be created nor destroyed but can only change from one state or location to another. It is a mass balance of the flow and storage of water within the target area, using the hydrologic equation:

$$\text{Inflow} - \text{Outflow} = \text{Change in storage}$$

Since no significant amounts of water are decomposed or composed within the system, the water content of the system (e.g., soil profile) can not increase without addition from the outside, nor can it diminish unless transported to the atmosphere by evapotranspiration or to deeper zones by drainage/recharge.

From mass balance, we can write:

$$P + Q_{in} = Q_{out} + ET \text{ (or E)} + \Delta S \tag{3.2}$$

Where, P is Precipitation /Rainfall; Q_{in} and Q_{out} are water flow into and out of the site, respectively; ET is evapotranspiration; E is the evaporation; and ΔS is change in water storage.

All components must be given as rates (depth per unit time (e.g., mm/day or mm/year) or, volume per unit time).

Individual components may consist of subcomponents. Re-writing the water-budget equation to incorporate many of the subcomponents results in:

$$P + (Q_{in\text{-}sw} + Q_{in\text{-}gw}) = (R_o + Q_{out\text{-}gw} + Q_{out\text{-}bf}) + (ET_{sw} + ET_{pw} + ET_{gw}) + (\Delta S_{sw} + \Delta S_{gw} + \Delta S_{pw} + \Delta S_{snow}) \quad (3.3)$$

Where,

Q_{in} = water flow into the site, where subscript *sw* and *gw* represents surface water and groundwater, respectively

ET = evapotranspiration, where subscripts *sw*, *pw* and *gw* represent surface water, profile water and groundwater, respectively

O_{out} = water flow out of the site, where subscript *gw* and *bf* represents groundwater and base-flow, respectively

R_0 = Surface water flow from the site (runoff)

ΔS = Change in storage, where subscripts *sw*, *gw*, *pw*, *snow* represent surface water, groundwater, profile water, and snow, respectively

All quantities are expressed in the same unit (in terms of volume of water per unit area, or equivalent depth units) during the period considered.

Groundwater recharge, R, includes any infiltrating water that reaches the saturated zone, and can be included in groundwater mass balance as:

$$R + Q_{in\text{-}gw} = Q_{out\text{-}gw} + Q_{bf} + ET_{gw} + \Delta S_{gw} \quad (3.4)$$

Where, $Q_{in\text{-}gw}$ represent all incoming sources to groundwater except recharge.

The above equation states that all water arriving at the water table either flows out of the basin as groundwater flow, discharged to the surface water body as baseflow, evapotranspired, or is retained in storage.

The above equation can be re-written as:

$$R = Q_{out\text{-}gw} - Q_{in\text{-}gw} + Q_{bf} + ET_{gw} + \Delta S_{gw} \quad (3.5)$$

Substituting this equation into Eq. (3.2) produces the following version of water budget:

$$R = (P + Q_{in\text{-}sw}) - R_0 - (ET_{sw} + ET_{pw}) - (\Delta S_{sw} + \Delta S_{pw} + \Delta S_{snow}) \quad (3.6)$$

For a particular basin (or site), some of the terms of the above water budget equation (eqn.3.5) may be negligible, and hence may be ignored.

Many hydrological models use the water budget equation. Estimation /determination method for each component is described in a later section (*section-3.13*).

Merits of Water Budget Method

(i) Simple to understand
(ii) Easy to apply, if other components can be measured/estimated accurately

(iii) Can be applied over a wide range of space and time scales, ranging from lysimeters (centimeters, seconds) to global climate models

(iv) For some areas/basins, data of water balance components (e.g., rainfall, runoff, water levels) may be readily available

Demerits of Water Budget Method

(i) An indirect method

(ii) Not easy to measure all the components accurately

(iii) Recharge is the residual or remainder of all other hydrologic components in the water-balance equation, thus errors can be high, with the errors in all the other fluxes accumulating in the recharge estimate.

(iv) Accuracy of recharge estimate depends on accuracy with which other components are measured.

The third limitation [No. (iii)] is critical when the magnitude of the recharge rate is small relative to that of the other variables, in particular ET and surface runoff. In this case, small inaccuracies in values of those variables commonly result in large uncertainties in the recharge rate. Some researchers have therefore questioned the usefulness of water-budget methods in arid and semiarid regions. However, if the water budget is calculated on a daily time step, precipitation sometimes greatly exceeds ET on a single day, even in arid settings. Averaging over longer time periods tends to dampen out extreme precipitation events (those most responsible for recharge events).

3.7.2. Water-Table Fluctuation Method

Quantification of natural recharge may be done from groundwater fluctuation or rise in water level (or piezometric level). This is also an indirect method.

Principle of the Method

In this approach, the rise in water level (i.e., refill) is assumed to caused by recharge. The refill depth is divided by the specific yield of the aquifer to get the recharge depth. Dividing the recharge depth by the time period gives the recharge rate.

Appropriateness/Suitability of the Method

To suit this approach, there should have a distinct rainy season with the remainder of the year being relatively dry.

Assumptions

The basic assumption in this approach is that the rise in the water-table is primarily due to the recharge. It is recognized that other factors such as pumping or sub-surface inflow-outflow during the rainy season do not exist or have no influence. It is also assumed that every inflow and outflow is uniformly distributed over the area.

Mathematical Formulation

Let,

The depth to water-table at the beginning of rainy season (say, at Julian day N_1) = d_1 m

And, the depth to water-table at the end of rainy season (say, at Julian day N_2) = d_2 m

Here, $d_1 > d_2$, because water-table will replenish/refill during this period by recharge.

Thus,

The difference in depth = $(d_1 - d_2)$ m

The difference in time = $(N_2 - N_1)$ days

Assume that, the specific yield of the aquifer = S_y $m^3/m^2/m$ (or % of aquifer volume)

Thus, actual difference in depth of water = $(d_1 - d_2) \times S_y$ m

And, the recharge rate = (difference in depth of water)/time involved

That is,

$$R = \frac{(d_1 - d_2) \times S_y}{(N_2 - N_1)} \text{ m/day} \tag{3.7}$$

Or,

$$R = \frac{S_y \times \Delta d}{\Delta t} \text{ m/day} \tag{3.8}$$

Where, Δd represents change in water-table depth (m), and Δt represents the time difference (days).

If pumping (abstraction) is done during the rainy season, the recharge rate can be computed as:

$$R = [S_y \times \Delta d + T_p]/\Delta t \tag{3.9}$$

Where, T_p is the pumping volume during rainy season (study period) divided by the influenced area (i.e., $T_p = Q/A$).

In this approach, the specific yield of the aquifer must be known. Water-level fluctuations occur in response to spatially averaged recharge. The area represented by the recharge rates can ranges from tens of square meters to several hundred or thousand square meters. Time periods represented by the recharge estimates can range from event scale to the length of the hydrographic record.

Specific yield (S_Y) can be estimated/determined in several ways, as:

(a) S_Y = Porosity – Specific retention (3.10)

(b) Specific yield can also be estimated from the difference in water content of the formation materials at saturation and field capacity (after gravity drainage). That is,

$$S_Y = W_{sat} - W_{FC} \tag{3.11}$$

Limitations

- The method is best suited for a seasonal or yearly estimate in regions having shallow to moderate water-table depth that display sharp rises and declines in water levels. Very short-term estimate may be associated with error due to time-lag of percolated water to reach the groundwater level, and measurement also must fall within the rising limb of water-table fluctuation pattern.
- Difficulties in applying the method are related to determining a representative value for specific yield and ensuring that fluctuations in water levels are due to recharge, but are not the result of pumping, or other phenomena, such as changes in atmospheric pressure, the presence of entrapped air, etc.
- The WTF method provides actual recharge (under the existing climatic and withdrawal conditions). If the groundwater withdrawal is not in potential rate, it does not provide potential estimate (as the aquifer may be full or the capacity may be limited).
- Suited for unconfined aquifer. For confined aquifer, complex calculation is required – with layer-wise specific yield from the level of piezometric surface to the aquifer.
- Hydrographs of monthly groundwater levels and rainfall revealed that groundwater level peak usually take place at a lag-time period (of about 3 – 4 months) with rainfall peak. This should be taken into account while observations are taken.
- The inherent assumptions may be approximately true for the rainfall and even for the return flow from irrigation but it is rarely true for the abstraction from the aquifer. When pumping is reduced or ceases during the rainy seasons, a redistribution of groundwater heads occurs so that part of the observed increase in water level may be due to normal well recovery. Moreover, the above equation is dependent on the value of the specific yield, which is difficult to determine since the water table fluctuation occurs in the partially saturated zone.

Merits of the Method

(i) Easy to understand and apply
(ii) Only one parameter need to be determined
(iii) Analysis of water-level fluctuations can, however, also be useful for determining the magnitude of long-term changes in recharge caused by climate or land-use change.
(iv) The method can be applied over a wide variety of climatic conditions.

Demerits of the Method

(i) An indirect method
(ii) Observation well should be installed and water-table data should be monitored
(iii) Accuracy of recharge estimates depends on the accuracy of specific yield

3.7.3. Lysimeter Method

Various components of 'soil water budget' can be accurately measured by using lysimeters. Lysimeters consist of containers filled with disturbed or undisturbed soil, with or without vegetation, and are hydrologically isolated from the surrounding soil, permitting determination of any term of the hydrologic equation (e.g., evapo-transpiration, percolation, etc.) when the others are known (Ali, 2010).

All lysimeters are designed to allow collection and measurement of drainage.

Lysimeters can be grouped into 2 main categories:

(a) Weighing lysimeter
(b) Non-weighing lysimeter

The weighing lysimeters have various weighing principles and devices. They may be based on weighing with varieties of scales and balances or on electronic weighing with strain gauge load cells, or a combination of both mechanical & electronic devices, or on hydraulic weighing systems. The Non-weighing lysimeters are also called 'drainage' or 'volumetric' or 'compensation' or 'pan' lysimeter.

Surface areas of lysimeters can range from 1.0 m^2 to 300 m^2 for large pan lysimeters; depths range from 1.0 to 20 m. If the base of the lysimeter is not deeper than the root zone, measured drainage fluxes overestimate aquifer recharge rates. Therefore, lysimeters are generally unsuitable for areas with deep-rooted vegetation. Recharge rates can be estimated at time scales from days to years. The minimum water flux that can be measured using a lysimeter depends on the accuracy of the drainage measurements and the surface area of the lysimeter. For large lysimeters (surface area ~100 m^2), recharge rates of about 1 mm/year can be resolved. The upper flux that can be measured depends on the design of the drainage system but should exceed drainage fluxes in most natural settings. Most lysimeters have a drainage-collection system that is open to the atmosphere, which creates a seepage-face lower-boundary condition. For thick unsaturated zones, this artifact causes different moisture and pressure-head profiles in the lysimeter relative to those in the adjacent undisturbed area. In an effort to minimize the influence of the bottom boundary, some lysimeters have been built with a porous plate on the bottom that is set at a prescribed pressure head.

Measurement Procedure

The simplified water balance at lysimeter scale can be written as:

$$P + I + SM_i = ET \text{ (or E)} + R + SM_f \quad (3.12)$$

Where,

P = Precipitation
I = Irrigation (if water added)
R = Recharge
ET = Evapo-transpiration
E = Evaporation (if the no crop/vegetation on the surface)
SM_i = Initial total soil moisture within the lysimeter soil
SM_f = Final total soil moisture within the lysimeter soil

Thus, the recharge (R) is:

$$R = P + I - ET \text{ (or E)} + (SM_i - SM_f) \quad (3.13)$$

For recharge study, measurement should be started from the beginning of the rainy period, if rainfall occurs during a part of the year. For the region having rainfall all over the periods of the year, measurement can be done for a certain period, or the whole year. Water application to the lysimeter surface should be provided whenever the soil moisture depletion or tension of the top soil layers reaches a certain level, to void cracking, which facilitates preferential flow through the wall.

Although daily observations of drainage, soil water content etc. can be made; water balance and evapotranspiration values are normally calculated on a weekly or 10-days basis.

ET Measurement

The evapotranspiration of a crop can be measured through another weighing lysimeter, from the difference in weights between two time periods. It can also be derived from field plot (specially designed, experimental plot) data using water balance equation at the plot level. It can also be estimated from crop and weather data (Smith et al., 1998).

Limitations

- Lysimeters are generally unsuitable for areas with deep-rooted vegetation.
- Bottom boundary does not represent actual field condition

Merits of Lysimeter Method

(i) Direct and precise method
(ii) Few other parameters to be known
(iii) If the soil column within lysimeter can be set similar to that of natural field soil, it gives representative estimates of recharge.

Demerits/Shortcomings

(i) Expensive
(ii) Not transportable
(iii) Difficult to construct, requires expert skill
(iv) Have high maintenance requirements.
(v) If the size of the lysimeter is small, side-effect (bypass flow *or* channeling *or* preferential flow) can cause excessive recharge
(vi) Soil profiling and density may not be exact as that of the field soil
(vii)Lysimeter soils can not represent spatial variability produced by natural and human-induced changes in surface- and subsurface-flow pathways.

3.7.4. Seepage Meter Method

Originally seepage meter was developed to measure canal seepage. Downward seepage (i.e., deep percolation, or recharge component) from surface-water bodies can be measured by using seepage meters. A seepage meter consists of a cylinder that is pushed into the bottom of the low-land, canal, stream or lake. Various types of seepage meters, including automated models, have been developed and are in use. Here, seepage meter with submerged flexible water bag is discussed. It is the simplest device regarding construction as well as operation. It consists of a water-tight seepage cup connected by a hose to a flexible (plastic) water bag floating on the water surface.

Principle of the Method

Attached to the cylinder is a reservoir of water; the rate at which water within the cylinder infiltrates (i.e., recharge) is determined by changes in reservoir volume.

Recharge rate (depth/time) = (Volume of water infiltrates)/(Internal area of the meter × time)

Detail Method

During measurement, the seepage meter is set under water. Water flows from the bag into the cup, where it seeps through the canal sub-grade area isolated by the cup. By keeping the water bag submerged, it will adapt itself to the shrinking volume so that the heads on the areas within and outside the cup are equal. The recharge rate is computed from the weight (and then converted to volume) of water lost in a known period of time and the area covered by the meter, i.e.,

$$R = V/(t \times A) \tag{3.14}$$

Where,
R = Recharge rate ($m^3/m^2/hr$)
V = Volume of water lost (m^3)
t = Time period (hr)
A = Area covered by the meter (m^2)

Uncertainties in estimated fluxes can be determined from replicate measurements. Time scales range from those based on individual events to days. Recharge over longer times is estimated from a summation of shorter times. Multiple measurements and longer time period can give confidence to the recharge rate measured.

Merits

(i) Direct method
(ii) Gives a rapid measurement
(iii) Relatively cheap
(iv) Easy to transport and apply the instrument in field
(v) Easy to calculate the recharge

Demerits

(i) Because seepage meters provide point estimates of water fluxes, measurements may be required at many locations for a representative value to be obtained.

3.7.5. Field Plot Water Balance

Recharge rate can be estimated as the difference between water added to the field plot system (net irrigation and rainfall) and evapotranspiration.

Principle

The principle in this method is that the water loss from a crop field takes place as evapotranspiration, surface runoff, and seepage and deep percolation (recharge). That is,

$$R = P + I - SD - \Delta SM - ET_c \tag{3.15}$$

Where, R is the deep drainage or recharge, P is the rainfall, SD is the surface drainage, ΔSM is the change in soil water storage (i.e., SM_i - SM_f), and ET_c is the crop evapotranspiration (or evaporation for bare field).

Detail Method

For ponded condition (particularly rice field), if no water is added to nor drained out from the field, then the total water use would be due to ET and seepage & percolation, and would be reflected by a corresponding subsidence or loss of head of water on the field surface. If we can control the seepage by inserting polythene around the experimental plot, or can make a buffer zone (say, 5 m wide) around the plot, then the horizontal flow of water from the plot *or* seepage can be controlled.

Assuming that daily evapo-transpiration is computed from weather data or pan evaporation data and subtracted from this head loss (fall in depth of water), the remainder of the daily fall in water level is equal to the recharge (as seepage is controlled).

For the measurement of fall of water-level, inclined meter sticks at a specified slope (say, 5:1) may be placed in the field supported with wooden frame. This inclination provides magnification of fall in water depth. Water level on the inclined meter scale may be recorded everyday at 9 A.M (or weekly, if the recharge rate is slow). For example, the meter scale is inclined at a slope of Z:1. The recharge rate is computed as:

$$R = \frac{WD_i - WD_f}{N{'}\ Z} - ET_t \tag{3.16}$$

where,

R = Recharge (mm/d)

WD_i = Water depth on the meter stick at the start of the measurement (Initial depth) (mm)

WD_f = Water depth on the meter stick at the end of the measurement (final depth) (mm)

ET_t = Evapotranspiration per day between the measurement periods (mm/d)

N = Number of days between the two measurement (d)

Z = Slope of the meter stick

If the meter stick is not inclined, i.e., vertical, the value of Z is one (1). If there is rainfall between the two measurement periods, the average daily rainfall amount should be subtracted from the above equation.

Merits of the Method

- Direct method
- Simple to understand

- No much skill of the operator is required

Demerits

- Water balance technique provides estimate of deep percolation (or recharge) for each individual irrigation (or rainfall). But in areas with heavy rainfall, it may give potentially greater error for the whole year (due to unreliable estimates of surface runoff).
- Accuracy of recharge estimates depends on accuracy of measuring of other component parameters.

3.7.6. Soil-Water Balance Approach

The soil-water balance approach accounts for soil water entering, leaving, and remaining within the root zone, and can be expressed in the following form:

$$R = (ASW + P) - (SRO + ET + SC) \tag{3.17}$$

Where,
R = deep percolation (i.e., potential recharge)
ASW = antecedent soil-water within the root zone
P = precipitation
SRO = surface runoff
ET = actual evapo-transpiration
SC = total soil-water storage capacity of the root zone.

All the above components are expressed in depth per unit time (e.g., mm/day, mm/month).

The soil-water balance approach emphasizes the physical factors (climate, soils, and vegetation) that determine the availability of water for recharge and consumptive water use. To implement this approach, one should compute the R (recharge) for different climatic stations/zones for the various possible combinations of the soils and vegetation types in the study area. Then, one should weigh the outputs on the basis of percentage of occurrence of the various vegetation types and soils within the grid elements in the study area.

Merits

- Easy to understand and apply
- The components can be measured/estimated with ease

Demerits

- Indirect method, recharge is the residual of the balance equation
- Accuracy of recharge estimate depends on the accuracy of the other components
- Only estimates recharge from soil matrix (Ignores the recharge from water-bodies, depressions)

- The method yields a point estimate (and then extrapolates to the whole study area).
- The ranges of recharge may be exaggerated or underestimated due to the finite number of the rainfall stations.

3.7.7. Zero-Flux Plane

Principle

The soil-water budget can be simplified by equating recharge to changes in soil-water storage below the zero-flux plane (ZFP), which represents the plane where the vertical hydraulic gradient is zero. The ZFP separates upward (ET) from downward (drainage *or* recharge) water movement. The rate of change in the storage term between successive measurements is assumed to be equal to the drainage rate to the water table or the *recharge rate*.

Methods

The ZFP requires soil matric-potential measurements to locate the position of the ZFP (Figure 3.1) and soil-water-content measurements to estimate storage changes. The position of the zero flux plane can be determined from the hydraulic profiles measured by the tensiometers. A transition zone can exist in the soil profile above which there is an upward soil-water flux (dH/dz > 0) and below which there is a downward flux or drainage (dH/dz < 0). The transitional zone is termed as zero flux plane (dH/dz = 0).

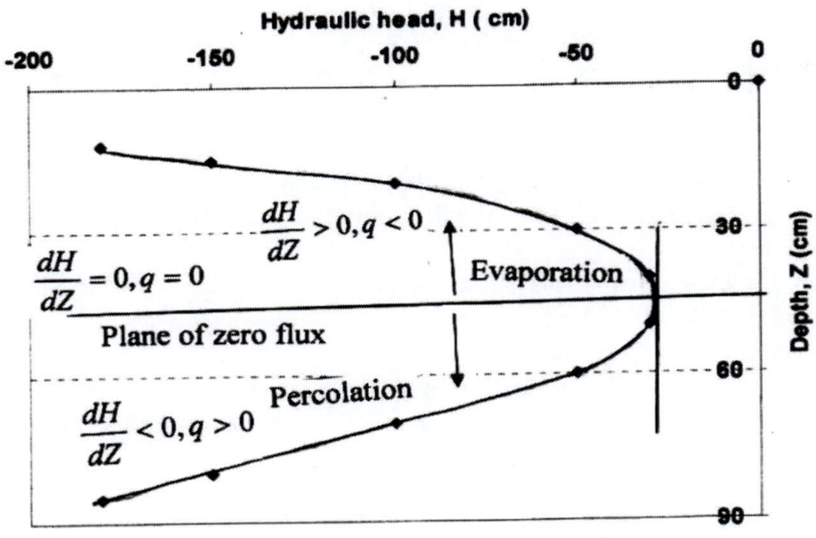

Figure 3.1. Schematic of zero flux plane in soil layer.

The average depth of the zero flux plane $\overline{z}_0$ is taken as the arithmetic mean during a particular time period. Soil moisture storage (S) at any time (t) is:

$$S = \int_{z_1}^{z_2} (\theta) dz \tag{3.18}$$

Recharge rate (R) is:

$$R = (S_1 - S_2)/(t_2 - t_1) \tag{3.19}$$

The minimum recharge rate that can be measured is controlled by the accuracy of the water-content measurements (generally ±0.01 m^3/m^3).

The ZFP provides a recharge estimate at the measurement point. Time scales range from event scales to years.

Fluxes and water balances can always be calculated using Darcy's equation, but the existence of a zero flux plane may make it possible to calculate soil-water balances without Darcy's equation. In situations without and with a zero flux plane we may proceed as follows:

In Absence of a Zero Flux Plane

Under this condition, there is either continuous drainage or continuous upward movement, and flux density is determined by Darcy's equation as:

$$q_z = K(\bar{q})\frac{dH}{dz} \tag{3.20}$$

where,

$K(\bar{\theta})$ = hydraulic conductivity corresponding to an average soil-water content $\bar{\theta}$ during the period considered

dH = difference in hydraulic potential

dz = difference in distance over which the difference in hydraulic potential is measured

dH/dz = average hydraulic potential gradient or the driving force

where: H = h +z

h = matrix potential or soil-water pressure head

z = gravitational potential or gravitational head

In some books, negative sign is imposed in Darcy's equation (in front of K) because the direction of the flux is where the potential decreases. Therefore, upward or downward flux in the profile will depend on the sign of the hydraulic gradient.

Limitations

The ZFP technique cannot be used when water fluxes are downward throughout the entire profile (periods of high infiltration) or when water storage is increasing, because downward movement of a wetting front generally masks the zero-flux plane.

Merits

(i) Direct method
(ii) This technique works best in regions where large fluctuations exist in soil-water content throughout the year and where the water table is always deeper than the ZFP.

Demerits

(i) The method yields a point estimate
(ii) Requires multiple readings at multiple locations for representative estimate
(iii) Data recording and calculation requires expertise
(iv) Accuracy of recharge estimate is controlled by the accuracy of the water-content measurements
(v) The ZFP technique is relatively expensive in terms of the required instruments and amount of data collection.

3.7.8. Darcy's Law Approach

3.7.8.1. Darcy's Law Method for Unsaturated Zone

Principle

If flow under field conditions is steady and driven by gravity alone, then, according to Darcy's Law, downward percolation rate will be numerically equal to the hydraulic conductivity of the material at the measured *in-situ* water content multiplied by hydraulic gradient.

Method

Darcy's law can be used to calculate recharge (R) in the unsaturated zone according to the following equation:

$$R = K(\theta)dH/dz = K(\theta)\frac{d}{dz}(h + z) = K(\theta)\left(\frac{dh}{dz} + 1\right) \quad (3.21)$$

Where,
$K(\theta)$ = the hydraulic conductivity at the ambient water content,
H = total head;
h = matric pressure head; and
z = elevation.

Application of Darcy's law requires measurements or estimates of the vertical total-head gradient and the unsaturated hydraulic conductivity at the ambient soil-water content. For thick unsaturated zones, below the zone of fluctuations related to climate, in uniform or thickly layered porous media, the matric pressure gradient is often nearly zero, and water movement is essentially gravity driven. Under these conditions, little error results by assuming that the total head gradient is equal to 1 (unit-gradient assumption). The unit-gradient assumption removes the need to measure the matric pressure gradient and sets recharge equal to the hydraulic conductivity at the ambient water content.

The minimum recharge rate that can be estimated by using Darcy's law depends on the accuracy of the hydraulic conductivity and head-gradient measurement if the latter is not unity. If hydraulic conductivity is strongly dependent on water content, uncertainty increases. The Darcian method does not necessarily indicate total recharge as it only accounts for diffuse, matrix flow. Recharge due to preferential flow is inherently non-Darcian and, if significant, must be determined separately.

Suitability/Limitations

- Not suitable in humid areas, where a lot of precipitation occurs.

Merits

(i) It can be applied throughout the entire year, whereas the ZFP can be applied only at certain times of the year.

Demerits

(i) This method provides a point estimate of recharge over a wide range of time scales; however, if applied at significant depths in thick vadose zones, it may represent a larger area.

(ii) Accuracy of recharge estimates depends on the accuracy of hydraulic conductivity and matric pressure head (or moisture content, θ) value. Unsaturated hydraulic conductivity is seldom measured at the exact field water content, therefore, interpolation of the measured, highly nonlinear $K(\theta)$ relationship in general is critical to recharge estimation.

3.7.8.2. Darcy's Law Method for Saturated Zone

Darcy's law can be used to estimate flow through a cross section of an unconfined or confined aquifer. This method assumes steady flow and no water extraction. The sub-surface water flux (q) is calculated by multiplying the hydraulic conductivity by the hydraulic gradient. The hydraulic gradient should be estimated along a flow path at right angles to potentiometric contours. The volumetric flux through a vertical cross section of an aquifer (A) is equated to the recharge rate (R) times the surface area that contributes to flow (S):

$$qA = RS$$

or,

$$K(\theta) \times (d\theta/dz) \times A = R \times S$$

or,

$$R = [K(\theta) \times (d\theta/dz) \times A\ ]/S \tag{3.22}$$

The cross section should be aligned with an equipotential line. This technique can be applied to large regions (~1 to ≥10,000 km^2).

The Darcy method has been used by Theis (1937) and Belan and Matlock (1973). The area should reflect a natural system with minimal pumpage. The time periods represented by the recharge estimates range from years to hundreds of years.

To apply this method, value of K, hydraulic gradient and S should be known. The applicability of laboratory-measured hydraulic conductivities at the field scale is not appropriate.

Merits

- The method is easy to apply if information on large-scale, effective hydraulic conductivity and the hydraulic gradient is available.

Demerits/Shortcomings

- Recharge estimates based on Darcy's law are highly uncertain because of the high variability of hydraulic conductivity (several orders of magnitude) over large area (even within a field).

3.7.9. Base-Flow Discharge

In watersheds with gaining streams, groundwater recharge can be estimated from stream hydrograph separation, i.e., base-flow.

Principle of the Method

Use of base-flow discharge to estimate recharge is based on the water-budget approach, in which recharge is equated to discharge. That is, Recharge = Base-flow.

This approach assumes that pumpage, ET and under-flow (from the aquifer) are negligible.

Where pumpage, evapotranspiration, and underflow to deep aquifers are significant, recharge (R) can be estimated as (assuming no change in groundwater level):

$$R = \text{Pumpage} + \text{ET} + \text{Under-flow} + \text{base-flow} \tag{3.23}$$

These other discharge components should be estimated independently.

Detail Method

Hydrograph analysis is an essential step of this method. Bank storage may complicate hydrograph analysis because water discharging from bank storage is generally derived from short-term fluctuations in surface-water flow and not from areal aquifer recharge and could result in over-estimation of recharge. Various approaches are used for hydrograph separation, including digital filtering (Nathan and McMahon 1990; Arnold et al. 1995) and recession-curve displacement methods (Rorabough 1964).

Recent progress has been made on the use of chemical and isotopic techniques to infer the sources of stream-flow from end members such as rainfall, soil water, groundwater, and bank storage (Hooper et al. 1990; Christophersen and Hooper 1992). This approach is data intensive, but it provides information that is useful in conducting hydrograph separation.

Merits

- Simple
- No sophisticated instrument is needed
- Recharge over longer times can be estimated by summation of estimates over shorter times.

Demerits

- An indirect method
- The accuracy of the estimated recharge rates depends on the validity of the various assumptions.
- Problematic to apply this method for large basins because of difficulties in separating surface-water and groundwater flow and bank-storage effects in larger systems and because of the areally uniform recharge assumption.
- Not suitable for basins having significant pumpage, evapotranspiration, and underflow to deep aquifers.
- The minimum time scale is a few months.
- May not represent potential recharge for all cases (i.e., where pumpage and ET are absent), but actual recharge

3.7.10. Numerical Method

3.7.10.1. Numerical Method for Watershed modeling

Watershed (rainfall-runoff) modeling is used to estimate recharge rates over large areas. Watershed models generally provide recharge estimates as a residual term in the water-budget equation (as discussed earlier).

The minimum recharge rate that can be estimated is controlled by the accuracy with which the various parameters in the water budget can be measured and the time scale considered. The various watershed models differ in spatial resolution of the recharge estimates. Some models are termed lumped and provide a single recharge estimate for the entire catchment. Others are spatially disaggregated into hydrologic-response units (HRUs) or hydro-geomorphological units (HGUs).

Merits

- Watershed models can be applied at a variety of scales. Small-scale applications allow more precise methods to be used to measure or estimate individual parameters of the water-budget equation.
- Time scales in models are daily, monthly, or yearly. Daily time steps are desirable for estimation of recharge because recharge is generally a larger component of the water budget at smaller time scales.

Demerits

- Indirect method, recharge is estimated as residual term
- Accuracy of recharge estimates is controlled by the accuracy with which the various parameters in the water budget can be measured and the time scale considered.

Bauer and Mastin (1997) applied the Deep Percolation Model to three small watersheds (average size 0.4 km^2) in the Puget Sound in Washington, USA. Average annual recharge rates were 37, 138, and 172 mm for the three basins. Arnold et al. (2000) applied the SWAT model to the upper Mississippi River Basin, USA (492,000 km^2). The basin was divided into 131 hydrologic-response units with an average area of 3,750 km^2. Estimated annual recharge ranged from ~10 to 400 mm. Other recent applications of watershed models to estimate recharge include Sami and Hughes (1996; recharge rate ~6 mm/year in a fractured system, South Africa), and Flint et al. (2002; recharge rate 2.9 mm/year at Yucca Mountain, Nevada, USA).

3.7.10.2. Numerical Modeling for Unsaturated-Zone Studies

Unsaturated-zone techniques for estimating recharge are applied mostly in semiarid and arid regions, where the unsaturated zone is generally thick. The recharge estimates generally apply to smaller spatial scales than those calculated from surface-water or groundwater approaches. Unsaturated-zone techniques provide estimates of potential recharge based on drainage rates below the root zone; however, in some cases, drainage is diverted laterally and does not reach the water table. In addition, drainage rates in thick unsaturated zones do not always reflect current recharge rates at the water table.

Unsaturated-zone modeling is used to estimate deep drainage below the root zone or recharge in response to meteorological forcing. Recent advances in computer technology and in computer codes have made long-term simulations of recharge more feasible. Theoretically, the range of recharge rates that can be estimated using numerical modeling is infinite; however, the reliability of these estimates should be checked against field information such as lysimeter data, tracers, water content, and temperature.

A variety of approaches is used to simulate unsaturated flow, including soil-water storage-routing approaches, quasi-analytical approaches (Simmons and Meyer 2000), and numerical solutions to the Richards equation. Examples of codes that use the Richards equation include HYDRUS-1D, HYDRUS-2D (Simunek et al. 1996), SWIM (Ross 1990), VS2DT (Hsieh et al. 2000), and UNSATH (Fayer 2000). Bucket-type models can be used over large areas (Flint et al. 2002); however, models based on the Richards equation are often restricted to evaluating small areas ($\leq$100 m^2) or to one-dimensional flow in the shallow subsurface ($\leq$15 m depth). Many recharge modeling studies evaluate periods of 30 to 100 years because of availability of meteorological information.

Merits

- Fast, with the aid of computer
- Time scales that can be evaluated range from hours to decades
- Numerical modeling can be used as a tool to evaluate flow processes and to assess sensitivity of model output to various parameters, and they allow predictions of future recharge regimes resulting from different land-uses and climatic changes

Demerits

- Because of uncertainties in hydraulic conductivity and non-linear relationships between hydraulic conductivity and matric potential or water content, recharge estimates based on unsaturated-zone modeling that use the Richards equation may be highly uncertain.
- Models based on the Richards equation are often restricted to evaluate small areas or to one-dimensional flow in the shallow subsurface.

3.7.10.3. Numerical Model for Saturated-Unsaturated Flow

Some numerical model may estimate recharge through saturated-unsaturated flow modeling through soil profile. The principal advantages of these types of numerical methods are that they attempt to represent the actual physical processes of interest and that they allow predictions of future recharge regimes resulting from different land uses and climatic changes. These advantages are often countered by the need to make simplifying assumptions in order to reduce the computational effort. For example, numerical models of the soil zone usually assume a single porosity medium with no spatial variation in properties. In practice many soils may have dual porosity, with preferred pathways during high saturation—that is, at times of recharge.

Merits

- They attempt to represent the actual physical processes of interest
- Can predict future recharge scenario resulting from land uses and climatic changes

Demerits

- Large data requirements often make application of numerical models difficult.

3.7.11. Tracer Techniques

Tracers may be natural or environmental and applied ones. Environmental tracers are those, which are available in the environment, but may differ in concentration/age with locations. Applied tracers are those, which are applied artificially, and include chemical tracers, stable isotopes, and radio-isotopic (or radioactive) tracers. The environmental tracers most commonly used in recharge studies include ^{2}H, ^{18}O, ^{36}Cl, ^{14}C, ^{13}C, and Cl.

Characteristics of an Ideal Tracer

An ideal tracer should have the following characteristics:

- The tracer needs to be mobile and soluble
- It should not be strongly retarded by the soil or aquifer matrix
- The tracer should be non-reactive and not transform during transport
- The tracer needs to be easily measured and easily extracted from the soil
- For artificial tracers, additional constraints need to be satisfied, such as low natural levels of the tracer in the environment, low toxicity, and low radioactivity.

- For environmental tracers, it is desirable to have large natural variations of tracer concentrations in the landscape. These constraints usually mean that only anions (Cl, Br, ^{36}Cl) or isotopically labeled water molecules (^{2}H, ^{18}O, and ^{3}H) can be used.

3.7.11.1. Chemical Tracer

Perspectives and Procedure

Chemical tracers are applied as a pulse at the soil surface, or at some depth within the soil profile, to estimate recharge. Infiltration of precipitation or irrigation transports the tracer to depth. Commonly used chemical tracers include bromide, and visible dyes. Organic dyes are generally used to evaluate preferential flow. The subsurface distribution of applied tracers is determined sometime after the application by digging a trench for visual inspection and sampling, or by drilling test holes for sampling. The vertical distribution of tracer (Figure 3.2) is used to estimate the velocity (v) and the recharge rate (R):

$$R = v\theta = \frac{\Delta z}{\Delta t}\theta \tag{3.24}$$

Where,

Δz = depth of the tracer peak,

Δt = time between tracer application and sampling, and

θ = volumetric water content

Tracers are generally applied at a point, or over small areas (10 to 200 m^2). The calculated recharge rates represent the time between application and sampling, which is generally months to years.

Different approaches/opinions are available for calculating recharge rate from concentration profile of the tracers, and are discussed in a later section.

The minimum water flux that can be measured with applied tracers depends on the time between application and sampling, and in the case of surface-applied tracers, the root-zone depth. If the rooting depth is assumed to be 0.5 m and the average water content is 0.2 m^3/m^3, a recharge rate of 100 mm/year would be required to transport the applied tracer through the root zone in 1 year. Lower recharge rates can be measured when the tracers are applied below the root zone. The maximum water flux that can be measured depends on the depth to the water table.

Merits

- Easy to apply and sampling
- No environmental hazard
- Cheap to buy
- Visual observation is possible for visible dye

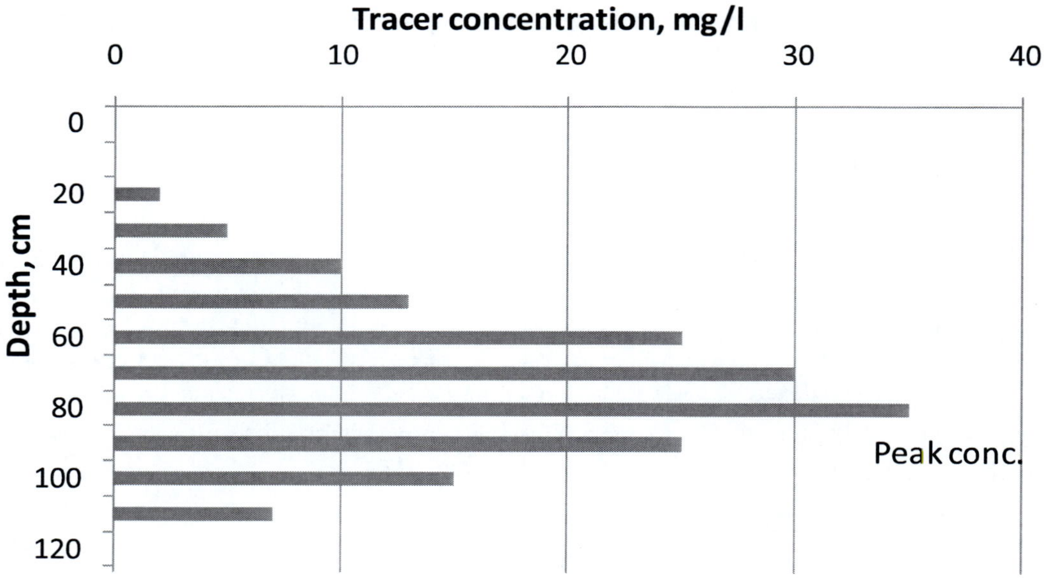

Figure 3.2. Schematic of concentration profile of chemical tracer.

Demerits

- Preferential flow may results in higher recharge rate
- For some chemicals, uptake by plants is often significant
- Sorption may be significant for organic dyes
- Concentration towards greater depth becomes negligible if the initial concentration/ amount is not high enough.

3.7.11.2. Isotopic Tracer

In many cases conventional hydro-geochemical studies are not sufficient to characterize groundwater hydrodynamics or to detect recharge areas and source areas of recharged water. The application of isotope-based methods has become well established for water-resource assessment, development and management in the hydrological sciences, and is now an integral part of many water-quality and environmental studies.

Stable Isotope

Stable isotopes of oxygen (^{18}O) and hydrogen (^{1}H, ^{2}H) are used to identify groundwater recharge from rivers and lakes. Since the isotopic composition of O and H in groundwater does not change as a result of rock–water interactions at low temperatures, it provides a useful means to detect source of recharge. In regions where rivers have headwaters at high elevations (such as in mountainous areas), river water is often depleted in stable isotopes relative to local precipitation in adjacent basins. If rivers retain the depleted isotopic signature of the headwaters, the difference in stable-isotope signatures of rivers and local precipitation can be used to determine the relative contribution of these two sources of groundwater recharge.

Stable isotopic tracers provide information on recharge sources; however, it is generally difficult to quantify recharge rates. The time scales range from seasonal in areas of high flux to hundreds of years in areas of low flux.

Radioactive Isotope

Radioactive isotopic tracers are applied as a pulse at the soil surface or at some depth within the soil profile to estimate recharge. Infiltration of precipitation or irrigation transports the tracer to depth. Commonly used radioactive tracers include tritium (^3H). The subsurface distribution of applied tracers is determined sometime after the application (3 months to 1 year) by drilling test holes for sampling. The vertical distribution of tracers (Figure 3.3) is used to estimate the velocity (v) and the recharge rate (R):

$$R = v\theta = \frac{\Delta z}{\Delta t}\theta$$

Where,

Δz = depth of the tracer peak,

Δt = time between tracer application and sampling, and

θ = volumetric water content.

Detail Working Method

1. Select a site (or multiple sites) considering factors such as level topography; representative soil type, coverage and drainage pattern; representative geo-hydrological condition; and accessibility. It should be far from electric pole, building, etc.
2. Choose the time period for tracer injection, preferably before the onset of monsoon (rainy season), or part of monsoon (depending on availability of study period).
3. Prepare the tracer solution (with low activity concentration – say, for tritium, activity 10-20 μC/ml).
4. In order to inject the tracer below the shallow root zone (also to avoid radiation hazard), make a hole with 1.0 – 2.0 cm diameter, 30~50 cm deep (may be with a specially designed drive rod, or with a small diameter auger).
5. Inject a small amount of tracer solution (2.5 – 10 ml) within the hole at the targeted depth (by a syringe, or a specially arrange device)
6. Fill back the hole with local soil.
7. Allow the setting up to the end of monsoon or the targeted time period.
8. At the end of time period, collect soil sample to a reasonable depth (1.5 – 3.5 m), depending on the time period between injection and sampling.
   Sampling can be done by soil coring auger with intact coring facility, or other suitable specially designed samplers. To have a good resolution for precise measurement, 10 - 15 cm sections are preferred.
9. Weigh the samples immediately after collection in the field by an electronic balance (having accuracy at least up to 0.05 gm), and double-seal in polythene/plastic bag.
10. Extract soil moisture from each sample in the laboratory by a vacuum distillation.
11. Determine soil moisture of each sample (by Torsion balance (Infrared light) or other suitable technique).

12. Take about 3-5 ml distillate (extracted soil-water) and mix with 10 ml of insta-gel (a universal liquid scintillation cocktail for counting aqueous and non-aqueous samples) in a low potash glass vial.
13. Count the activity of the tracer in mixture using a specified counter (for example - for tritium, by Liquid Scintillation Spectrometer, or other countable instrument).
14. Tabulate or plot the tracer activity of every sections against the sample depth for the profile.
15. Calculate the displacement of the tracer as:

Displacement = depth of peak concentration (activity) – injection depth

16. Calculate the average moisture content (volume basis) of the depth between injection and peak concentration.
17. Calculate the velocity, recharge rate, total recharge (from the equation specified earlier, eqn.3..).

The minimum water flux that can be measured with applied tracers depends on the time between application and sampling and, in the case of surface-applied tracers, the root-zone depth.

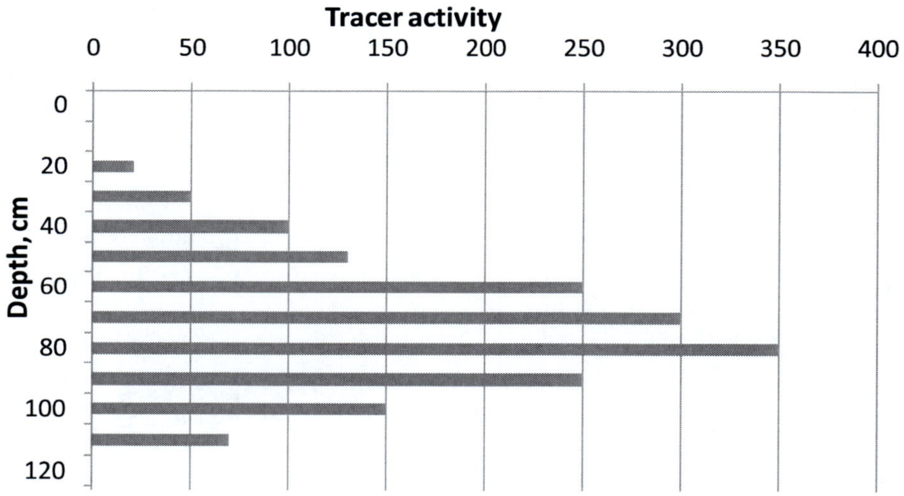

Figure 3.3. Schematic of activity profile of a radio-tracer.

Although ^{3}H is the most conservative of all tracers, its use is prohibited in many areas because of environmental-protection laws.

Merits

- Direct method
- More accurate results can be obtained
- No absorption or loss of tracer
- Generally require only one-time sampling and may represent long time periods.

Demerits

- Radioactive material, may not be permitted by environmental protection law in all areas
- Needs precaution
- Needs costly instrument for reading the sample (for ^3H concentration), and technical hand to operate the instrument
- Point estimates of recharge, therefore several measurements are needed to get the average/representative value.
- In areas having higher recharge rate, difficulties of soil sampling at higher depths and locating the tracer peak, specially for longer time period between injection and sampling.
- Water in the root zone can be evapotranspired, thus water content within the root zone can be underestimated.

Commonly used radioactive isotopes (also called radio-isotope) in hydrological studies and their half-life are given in Table 3.1. Half-life is the time period after which the radioactivity of an isotope is decayed to half of the original activity. Among the isotopes, Tritium and Carbon-14 have found the widest application in groundwater studies.

Table 3.1. Commonly used radioisotopes in hydrological studies and their half-life

Sl	Isotope	Half-life (year)
1	^{3}H (Tritium/Hydrogen-3)	12.43
2	^{14}C (Carbon-14)	5730
3	^{36}Cl (Chlorine-36)	3.01×10^5
4	^{129}I (Iodine-129)	16×10^6
5	^{39}Ar (Argon-39)	269
6	^{81}Kr (Crumium-81)	2.1×10^5
7	^{85}Kr (Crumium-85)	10.76

3.7.11.3. Environmental Tracers

Chloride Mass Balance (CMB) Approach

Perspectives and Methods

Environmental tracers such as chloride (Cl) are produced naturally in the Earth's atmosphere and are used to estimate recharge rates. The mass of Cl into the system (precipitation and dry fallout, P) times the Cl concentration in P (C_P) is balanced by the mass out of the system (drainage, D) times the Cl concentration in drainage water in the unsaturated zone (C_{uz}) if surface runoff is assumed to be zero. That is,

$$P \times C_p = D \times C_{uz}$$

Or,

$$D = \frac{P \times C_p}{C_{uz}} \tag{3.25}$$

Chloride concentrations generally increase through the root zone as a result of evapotranspiration and then remain constant below this depth. Drainage is inversely related to Cl concentration in the unsaturated-zone pore water (Eq. 3.25). This inverse relationship results in the Cl mass-balance (CMB) approach being much more accurate at low drainage rates, because Cl concentrations change markedly over small changes in drainage. The CMB approach has been most widely used for estimating low recharge rates, largely because of the lack of other suitable methods. The maximum water flux that can be estimated is based on uncertainties in measuring low Cl concentrations and potential problems with Cl contributions from other sources, and is generally considered to be about 300 mm/year.

Merits

- Easy to understand and apply
- Can also be applied in sandy areas

Demerits

- Is not suitable where surface runoff is present *or* significant
- Provides point estimates of recharge
- Uncertainties in measuring low Cl concentration
- Cl contributions from other sources (such as agricultural sites) may affect the result
- The accuracy of this approach decreases as recharge rates increase.

3.7.11.4. Historical Tracer

Perspectives and Procedure

Historical tracers result from human activities or events in the past, such as contaminant spills or atmospheric nuclear testing (^3H and ^{36}Cl). These historical tracers or event markers are used to estimate recharge rates.

Industrial and agricultural sources produce contaminants such as bromide, nitrate, atrazine, and arsenic, and these can provide qualitative evidence of recent recharge; however, uncertainties with respect to source location, concentration, and timing of contamination, as well as possible non-conservative behavior of contaminants, make it difficult to quantify recharge.

The presence of an event marker in water suggests that a component of that water recharged in a particular time period. The peak concentration of thermonuclear tracers can also be used to estimate water flux by using equation (3.24):

$$R = v\theta = \frac{\Delta z}{\Delta t}\theta$$

Where,

R = recharge rate

z = depth of the peak concentration of the tracer,

θ = average water content above the tracer peak, and

t = time period between the peak tracer fallout and the time the samples were collected.

Merits

- No cost of tracer
- No extra hazard

Demerits

- Uncertainties of the tracer (location and concentration)
- Water in the root zone can be evapotranspired, thus water fluxes estimated from tracers within the root zone can overestimate water fluxes below the root zone.
- The maximum water flux that can be estimated may be limited by depth to groundwater.
- In areas having higher recharge rate, difficulties of soil sampling at higher depths and locating the tracer peak.

Deep penetration of thermonuclear tracers has been found in sandy soils in arid settings (^3H, 23 mm/year, Dincer et al. 1974; ^3H, 22 to 26 mm/year, Aranyossy and Gaye 1992). The maximum water flux that can be estimated may be limited by depth to groundwater. For example, if the average water content is 0.1 m^3/m^3 and the time since peak fallout is 40 years, a recharge rate of 50 mm/year would result in a peak at a depth of 20 m. Therefore, this technique is generally unsuitable where recharge rates are much greater than 50 mm/year. Theoretically, the technique could be used for higher recharge rates if the water table were deeper; however, the difficulty of soil sampling at these depths and locating the tracer peak may be prohibitive.

Historical tracers or event markers such as bomb-pulse tritium (^3H) has been used widely used in the past in both unsaturated and saturated zones to estimate recharge. However, bomb-pulse ^3H concentrations have been greatly reduced as a result of radioactive decay. In the southern hemisphere, where ^3H concentrations in precipitation were an order of magnitude lower than in the northern hemisphere, it is often difficult to distinguish bomb-pulse ^3H from current ^3H concentrations in precipitation.

3.7.11.5. Groundwater Dating

Perspectives and Methods

Recharge rates can be determined by estimating ages of groundwater. Age is defined as the time since water entered the saturated zone. Groundwater ages can be estimated from CFCs by comparing CFC concentrations in groundwater with those in precipitation.

Age from $^3H/^3He$ Data

The age of the groundwater, t, can be calculated from $^3H/^3He$ data (ratio of tritium to tritiogenic helium) using the following equation:

$$t = -\frac{1}{\lambda} ln\left[1 + \frac{^3He_{trit}}{^3H}\right] \tag{3.26}$$

Where,
λ = decay constant (ln $2/t^{1/2}$), $t^{1/2}$ is the ^3He half life (12.43 years)
He_{trit} = tritiogenic ^3He

Use of this equation assumes that the system is closed (does not allow ^3He to escape) and is characterized by piston flow (no hydrodynamic dispersion).

Use of ^{14}C for Groundwater Age

Radioactive decay of ^{14}C can be used to estimate groundwater ages of 200 to 20,000 years. The estimated recharge rates are average rates over the time period represented by the groundwater age. The method is most accurate where piezometers have been completed with relatively short well screens. Recharge rates calculated using groundwater dating spatially integrate recharge over an area up-gradient from the measurement point.

The residence time of groundwater can be calculated through the following decay equation:

$$\ln(A/A_0) \times 8266.7 = \text{Age}$$

where A_0 is initial radiocarbon concentration of water.

The activity of ^{14}C may be low under certain circumstances. The reason for the low ^{14}C activity in groundwater may be due to the interaction of the infiltrated water and the carbonate rock (limestone) formation outcropping in the area. As the water infiltrates down to the water table, it can be considered under a closed-system condition, which means it is isolated from the influence of the CO_2 reservoir. The dissolution of carbonate by water–rock interactions lowers the ^{14}C activity of groundwater by substitution of "dead" carbon from the aquifer rock (Bajjali et al. 1997).

Recharge Rate from GW Age

Most of the recently developed isotope techniques are aimed at determining the age of water, which in turn permits calculation of groundwater travel time. The recharge rate, *R*, can then be calculated as:

$$R = \frac{L\varphi_e}{T} \tag{3.27}$$

where L is the distance along the flow path, $\varphi_e$is the effective porosity, and T is the travel time or age of the groundwater at the distance L.

Three basic types of groundwater dating methods are available. These are:

1. Those methods which rely on input concentrations that have changed in time and are well known, such as the radioactive noble gas krypton-85, and the synthetic organic compounds chlorofluorocarbons (CFCs), used for dating young waters (less than ~40 years old);
2. Tracers for which input concentrations have been constant, and decreases in concentration with time due to radioactive decay, such as ^{14}C, used for dating waters over the time scales of 200 to 20,000 years;

3. Methods where the input concentrations may have changed with time but can be determined because both parent and daughter isotopes are measured, such as $^{3}H/^{3}He$, which ratio also is used to date young waters (0 to ~50 years).

Characteristics and Considerations

In unconfined porous-media aquifers, groundwater ages increase with depth, the rate of which depends on aquifer geometry, porosity, and recharge rate (Cook and Bohlke 2000). The vertical groundwater velocity decreases with depth to zero at the lower boundary of the aquifer. The age increases linearly with depth near the water table and nonlinearly at greater depths. Near the water table, the influence of the aquifer geometry is greatly reduced. The recharge rate can be determined by dating water at several points in a vertical profile, calculating the groundwater velocity by inverting the age gradient, extrapolating the velocity to the water table if it is not measured near the water table (Cook and Solomon 1997), and multiplying the velocity by the porosity for the depth interval. That is,

$$t = v \times \rho, \text{ where } \rho \text{ is the porosity.}$$

CFCs and $^{3}H/^{3}He$ are used to determine groundwater ages up to approximately 50 years, with a precision of 2 to 3 years (Cook and Solomon1997).

Horizontal flow velocities can be estimated from radioactive decay of ^{14}C or ^{36}Cl in a confined aquifer. These data can be used to estimate recharge rates (R):

$$R = vnA/S \tag{3.28}$$

where v is velocity, n is porosity, A is the cross-sectional area of the confined aquifer where the velocity is determined, and S is the surface area of the recharge zone.

If necessary, corrections should be made for any leakage to or from the confined aquifer.

Merits

- No field setting/experiment is needed
- Easy to implement if the instrument for reading the sample is available

Demerits

- Costly instrument is needed for reading the sample
- The range of recharge rates that can be estimated by using groundwater dating depends on the ranges of ages that can be determined.
- Variation of isotopic signature with depth may occur (due to various reasons), so multiple sampling (throughout the depth up to aquifer) is needed.

3.7.11.6. Limitations/Restrictions of Using Tracer

There are various restrictions to the use of these tracers to date groundwater. CFCs can be used only in rural areas not affected by septic tanks because of contamination associated with industrial and residential areas, whereas $^{3}H/^{3He}$ can be used in contaminated and uncontaminated areas. The concentrations of $^{3}H/^{3}He$ and CFCs at the water table are assumed to be equal to those in the atmosphere. Any difference in concentrations would result in errors in the estimated ages. Cook and Solomon (1995) conclude that errors are negligible if the unsaturated-zone thickness is ≤10m. Other issues that need to be considered when using these tracers include the effect of excess air for both $^3H/^3He$ and CFCs and the effect of recharge temperature, sorption, and degradation on CFCs. Sampling for $^3H/^3He$ and CFCs is complex, and analysis is relatively expensive.

3.7.11.7. Interpretation of Tracer Results

Mechanisms of tracer infiltration (piston flow vs preferential flow) will affect the interpretation of results. Although piston flow is often able to explain the behavior of tracers in the field, but in humid regions water can move along preferred pathways. Thus, preferential or non-piston-type flow must be dealt with in any comprehensive analysis of recharge.

The following approaches have been used for estimating recharge rates from tracer concentration profiles in the unsaturated zone (Allison et al., 1994):

(i) *From the tracer peak*
In this approach, the peak of tracer concentration represents the travel distance (Δz). Any bypass (preferential) flow will result in recharge being under-estimated.

(ii) *From the shape of the tracer profile in the soil*
From the shape of the tracer profile, information about flow mechanisms can be obtained. In order to obtain estimates of mean annual recharge, a weighting function that takes into account year-to-year variations of recharge is needed.

(iii) *From the total amount of tracer*
In this approach, recharge is considered as the total amount of tracer (T_r) stored in the profile.

$$T_r = \int_0^z T(z)\theta(z)dz$$

where $T(z)$ is the tracer concentration of water in the unsaturated zone at a distance z beneath the surface, and $\theta(z)$ is the volumetric water content.

For evaporative tracers, such as ^3H, mean annual recharge (R_m) is estimated as:

$$R_m = \frac{T_r}{\sum_{i=1}^{z} W_i \times T_{pi} \times exp(-t\lambda)} \quad (3.29)$$

where w_i is the annual recharge weighting factors, T_{pi} is the tracer concentration of recharge water i years before the present; and λ is the tracer decay constant. In this analysis, non-piston flow can be handled because $R_m$is independent of the distribution of the tracer in the profile.

3.7.11.8. Merits of Tracer Techniques over Other Methods, and Concerns

Merits

- Movement of tracer is governed mainly by the long-term mean soil-water fluxes that lead to recharge.

- This technique does not require frequent visits to the field.
- Smaller fluxes can be estimated with this technique than with other methods.

This technique is used to compare/validate other techniques, and this is often the only alternative.

Concerns

As tracers do not measure water flow directly, a number of problems can arise, leading to over- or underestimation of recharge. These problems include secondary (unknown) tracer inputs, mixing, and dual flow mechanisms; such problems only arise if the sources, sinks, and pathways of tracer are not fully understood. Part of the recharge going through preferred pathways (such as root channels or fissures) may invalidate the results of a tracer study.

3.7.12. Empirical Method

In this approach, the recharge from a single rainfall event, or from monthly or yearly total rainfall, is estimated using empirical equation, which has been established previously for the specific basin *or* for the particular type of basin. The forms of the empirical equations may vary from basin to basin, but a common form of this type of equation is:

$$R = C_1 \times (P - C_2) \qquad (3.30)$$

where P is precipitation, and C_1 and C_2 are coefficients. For a particular basin *or* area (or based on the rainfall intensity and/or amount), the value of the coefficients may be constant. The theoretical hypothesis behind this type of form is that, above a threshold value of rainfall, a certain percentage of rainfall is contributed to groundwater through recharge.

The recharge increases with increasing precipitation, but recharge percentage is non-linear relationship with the precipitation.

Based on the studies undertaken by different scientists and organizations regarding correlation of groundwater level fluctuation and rainfall, some empirical relationships have been derived for computation of natural recharge to groundwater from rainfall. Such relationships are pertinent to the study area. Some of them are given below:

Anderson et al. (1992) Formula

Anderson et al. (1992) performed water budget analysis for series of basins in southern Arizona, and developed an empirical equation to estimate recharge on an annual time scale. The Anderson equation uses mean annual precipitation to compute mean annual recharge (R) as follows:

$$\log R = -1.40 + 0.98 \log(P-8) \qquad (3.31)$$

Where R is mean annual recharge and P is mean annual precipitation (all are in inches). Recharge will occur when precipitation is greater than 8 inches.

Chaturvedi Formula

Based on the water level fluctuations and rainfall amounts in Ganga-Yamuna doab (India), Chaturvedi in 1936, derived an empirical relationship to calculate recharge as a function of annual precipitation.

$$R = 2.0\ (P - 15)^{0.4} \tag{3.32}$$

where, R is the net recharge due to precipitation during the year (inch); and P is the annual precipitation (inch).

This formula was later modified by further work at the U.P. Irrigation Research Institute, Roorkee (India) and the modified form of the formula is:

$$R = 1.35\ (P - 14)^{0.5} \tag{3.33}$$

The Chaturvedi formula has been widely used for preliminary estimations of groundwater recharge due to rainfall. It may be noted that there is a lower limit of the rainfall below which the recharge due to rainfall is zero. The percentage of rainfall recharged commences from zero at P = 14 inches, increases upto 18% at P = 28 inches, and again decreases. The lower limit of rainfall in the formula may account for the runoff, soil moisture deficit, interception and evaporation losses. These factors being site specific, one generalized formula may not be applicable to all the alluvial areas.

Kumar and Seethapathi Formula

Kumar and Seethapathi () derived the following empirical relationship (similar to Chaturvedi formula) by fitting the estimated values of rainfall recharge and the corresponding values of rainfall in the monsoon season through the non-linear regression technique.

$$R = 0.63\ (P - 15.28)^{0.76} \tag{3.34}$$

where,

R = Groundwater recharge from rainfall in monsoon season (inch);
P = Mean rainfall in monsoon season (inch).

Merits of Empirical Methods

- Simple and easy to apply
- Useful for making first-guess estimates of recharge
- Can be developed and use for monthly, seasonal and yearly estimates
- Can be developed for an area if data of rainfall and recharge are available for similar area (similarity in geo-hydrologic, land-use and ET)
- Includes recharge from all other possible sources (intermittent flow, permanent water bodies), as the recharge are compared with the change in water-table

Demerits

- Needs an empirical equation (first to develop)
- More uncertainty for very dry area
- Change in physiographic conditions can alter the coefficients

Future Refinement

Multiple linear regression (MLR) can be tried to correlate between recharge and potential explanatory variables (if can be quantified properly). At a broad scale, such as a multi-state region, increased systematic variation in soil, landscape, and climate variables will facilitate improved correlations with recharge, assuming that recharge can be estimated with reasonable certainty. These correlations can be exploited with empirical models to identify factors that influence recharge and, ultimately, to predict recharge in unsampled/new areas.

3.7.13. Application of Multiple Techniques

There are some uncertainties associated with each approach for estimating recharge, therefore, use of many different approaches is recommended to ascertain the recharge estimates. In many cases, different approaches complement each other and help refine the conceptual model of recharge processes. Ideally, as many different approaches as possible should be used to estimate recharge. Examples of multiple approaches include the use of various tracers in unsaturated zones (e.g., ^{36}Cl and ^{3}H) and saturated zones (CFC-11, CFC-12, $^{3}H/^{3}He$). Some studies have combined soil physics and environmental tracers and also numerical modeling. Techniques based on data from surface-water and unsaturated and saturated zones can also be combined. Inverse groundwater modeling may be combined with groundwater-age data on the basis of ^{3}H, $^{3}H/^{3}He$, and ^{14}C to provide regional estimates of groundwater recharge.

A methodology consisting of a combination of soil-moisture budget and water-table rise analyses is known as *hybrid water fluctuation method* (Sophocleous, 1991). This combination methodology, designed to minimize water-balance errors, was successfully applied in the central Kansas plains, which are characterized by semiarid to sub-humid climate, relatively flat terrain, and relatively shallow water table.

Some studies have used catchment-scale surface-water models to provide estimates of recharge to groundwater models; however, such an approach assumes that no time lag occurs between infiltration and groundwater recharge. Surface-water and groundwater models have been integrated, such as the SWAT and MODFLOW codes by Sophocleous and Perkins (2000). This integrated model provides a framework for the total system that can be used to check continuity and better constrain model parameters. Parameter optimization is conducted by calibrating against multiple targets, such as groundwater levels, stream-flow data, and other data, that should result in more reliable results than obtained when using watershed or groundwater models separately. In this integrated model, recharge is constrained by an overall water budget for the surface-water system, and stream–aquifer interactions are constrained by the watershed model.

In summary, many techniques are restricted to measuring recharge rates within a certain range, which may not be known a priori before the recharge study is undertaken. Therefore, only different approaches that are likely to provide the most quantitative estimates of recharge can be suggested. Results provided by initial studies should be used as platforms for additional data to optimize the techniques and refine the recharge estimates. An iterative approach will be required to accurately quantify recharge rates, and a variety of approaches should be applied because of uncertainties in recharge estimates. Results from the various techniques can be compared to determine uncertainties in recharge estimates.

3.8. Recharge Estimation Related to Aquifer Vulnerability to Contamination

Although spatial variability in recharge at local and intermediate scales may not be critical for water-resource assessment, it is critical for contaminant transport, because focused recharge and preferential flow allow contaminants to migrate rapidly through the unsaturated zone to underlying aquifers. Locating areas of low recharge is important for radioactive and hazardous-waste disposal sites. Aquifer vulnerability to contamination is controlled to a large extent by recharge-transporting contaminants to aquifers; therefore, delineation of zones of high recharge is critical to defining zones that are vulnerable to contamination.

The time scales of recharge estimates are also important. Information on recharge at decadal time scales is generally required for water-resource planning, whereas time scales required for contaminant transport range from days to thousands of years, depending on the particular contaminants being considered.

Spatial variability in recharge and preferential flow through the unsaturated zone are critical issues for contaminant transport and for assessing aquifer vulnerability to contamination. The occurrence of focused flow and preferential flow should be identified because they greatly reduce transport times through the unsaturated zone. Quantifying recharge in areas subjected to preferential flow is quite complicated. Whereas preferential flow may be demonstrated using unsaturated-zone tracer experiments, recharge rates in areas of preferential flow should be based on saturated-zone data because the saturated zone generally integrates contributions from preferred pathways.

Preferential flow includes macropore flow and unstable flow. Macropore flow refers to flow along noncapillary-size openings such as those in fractured chalk, fractured basalts, karst, and structured fine-grained sediments. Macropore flow occurs mostly in humid sites that have high precipitation, or in arid settings subjected to ponding, or in cracking clays.

The importance of preferential flow in assessing aquifer vulnerability to contamination depends on the type of contaminants. Preferential flow is much more important for contaminants that exceed health standards in the parts-per-billion range (e.g., pesticides) than the parts per-million range (e.g., nitrate). Nitrate contamination requires movement of the bulk of the pore water, which is much greater than the generally smaller water volume transported along preferred pathways.

Estimation of recharge in areas of preferential flow is difficult. Preferential flow may be evidenced by multiple peaks of thermonuclear tracers such as ^{36}Cl and ^{3}H in the unsaturated zone. For calculating recharge rates, the tracer center of mass rather than the tracer peak should be used, because the latter does not conserve mass.

3.9. CHOOSING AN APPROPRIATE METHOD FOR RECHARGE ESTIMATION

3.9.1. Factors to be Considered in Selecting a Recharge Estimation Method

The climate, geomorphology (including topography, soil, and vegetation), and geology of a site control the location and timing of recharge, and therefore impact on the choice of a technique for estimating recharge. In general, the following factors affect the choice of a recharge estimation method:

(i) Aim *or* objective of recharge estimation
(ii) Required accuracy of recharge estimate
(iii) Geomorphology of the target area
(iv) Climate
(v) Geology
(vi) Source and mechanism of recharge
(vii)Temporal and spatial scale required
(viii)Availability of time and money
(xi) Limitations/suitability of the methods themselves

Aim or Objective of Recharge Estimation

The goal of the recharge study plays an important part in determining appropriate techniques for quantifying recharge because the objective determines the space/time scales required. Evaluation of water resources generally requires techniques that provide regional estimates of recharge, whereas spatial variability of recharge and preferential flow are critical to contaminant transport or determination of aquifer vulnerability to contamination.

Required Accuracy of Recharge Estimate

The required accuracy and reliability of recharge estimates influence the choice of technique. Saturated-zone techniques generally provide recharge estimates that are more reliable because they estimate actual recharge, whereas surface-water and unsaturated-zone techniques estimate potential recharge. With the exception of lysimeters, water-budget approaches are generally less accurate in semiarid and arid regions than in humid regions, because in dry areas recharge constitutes a smaller fraction of the water budget and the recharge term accumulates the errors in all the other terms of the water-budget equation.

Geomorphology of the Target Area

Geomorphology of the target area along with climate dictates the recharge rate, and hence the recharge estimation method.

Climate

Humid and arid systems represent end members for different climates and generally require different approaches to quantify recharge. Humid regions are usually characterized by shallow water-tables and gaining streams. Aquifers are often full, and groundwater is usually discharged through evapotranspiration and base-flow to streams. Diffuse recharge is dominant. Recharge rates in these regions are often limited by the ability of aquifers to store and transmit water, which depend on subsurface geology. In contrast, deep water-tables and losing streams are common in alluvial valleys in arid regions. Focused recharge commonly dominates. Recharge rates are limited in large part by the availability of water at the land surface, which is controlled by climatic factors, such as precipitation and evapotranspiration, and by surface geomorphic features.

Geology

An understanding of the physiographic provinces is useful in developing a conceptual model of recharge in a system, and analogies can be drawn to similar physiographic provinces elsewhere. Insight is thus gained into recharge sources, flow mechanisms, and spatial and temporal variability in recharge.

Alluvial fans near the base of mountain fronts usually receive recharge from streams draining adjacent mountains; therefore, techniques that quantify surface-water input to these systems are used to assess the total volume of recharge in alluvial fans.

Source and Mechanism of Recharge

Sources and mechanisms of recharge may also dictate the techniques to be used to quantify recharge. Surface-water sources require techniques such as channel-water budgets, heat tracers, and water-table fluctuations. Climatic regions, e.g., arid vs. humid, have fundamental differences in recharge that may require different approaches. Surface-water and saturated-zone techniques are more widely used in humid regions, whereas unsaturated-zone techniques are widely used in arid and semiarid regions. Watershed-modeling approaches may be more accurate in humid regions, where perennial surface-water flow can be used for model calibration. Although historical tracers can be used in the unsaturated zone in humid regions, their use is limited because of generally thin unsaturated zones and the ease of using such tracers in the saturated zone. Water-table fluctuations and Darcy's law could also be used in arid and semiarid regions where water-tables are shallow.

The estimated recharge rate at a site may determine the most appropriate procedures for quantifying recharge because different techniques measure recharge over different ranges. Initial estimates of recharge rates can be obtained by analogy to other sites having similar climatic and geomorphic settings or by numerical simulations that use existing climatic data. Water-table depth may also be important because some techniques (e.g., $^{3}H/^{3}He$, CFCs) are used only in areas of relatively shallow ($\leq$10 m) water-tables.

Temporal and Spatial Scale Required

A key factor in deciding on a recharge-estimation methodology is related to the spatial and temporal scale of interest. The space and time scales of the various techniques also affect the choice of technique to be used. Surface-water and groundwater approaches provide regional estimates of recharge, whereas unsaturated-zone techniques generally provide estimates at points or small scales. However, point estimates may represent much larger areas. Time scales of different approaches are variable. Surface-water approaches provide recharge estimates on the time scale of events or longer by summing up events, whereas unsaturated-zone and saturated-zone tracers provide recharge estimates over long periods ($\leq$ thousands of years), which may be important for studies related to disposal of radioactive waste. Long-term recharge estimates provided by tracers are not always advantageous, because they do not provide detailed information on variations in recharge over time. Unsaturated-zone techniques, such as lysimeters, zero-flux plane, and applied tracers, and saturated-zone techniques, such as water-table fluctuations, provide recharge estimates on event time scales also. These techniques are restricted to providing recharge estimates for the length of the monitoring record. In arid and semiarid areas where deep drainage fluxes are low and water-tables are deep, interpreting groundwater hydrographs and water-table rises may be misleading for estimating rates of groundwater recharge; chemical and isotopic methods are likely to be more successful than physical methods in such cases.

Availability of Time and Money

Additional concerns that can influence the choice of technique include time and expense constraints. Cost and time requirement for the various approaches generally vary considerably. If recharge estimates have to be developed within a short time (months), then techniques based on long-term monitoring (several years) cannot be used. In that case, tracer techniques may be more suitable. Although sampling and analysis of chemical and isotopic tracers are usually considered to be expensive, one-time sampling is generally sufficient; therefore, the costs may be less than those associated with long-term monitoring that require monitoring equipment and continual collection and analysis of data.

Limitations/Suitability of the Methods themselves

Surface-water techniques may be appropriate for quantifying recharge from streams using channel-water budgets (differential stream-flow measurements) or other techniques such as seepage meters. Techniques that require hydraulic-conductivity data, such as Darcy methods and unsaturated- and saturated-zone models, are inherently inaccurate because hydraulic conductivity can vary over several orders of magnitude. Various sources of uncertainty include those related to measurement of hydraulic conductivity, applicability of data at the measurement scale (laboratory vs. field scale) to the scale of recharge calculation, and spatial variability in hydraulic conductivity. Uncertainties in hydraulic conductivity are even greater in unsaturated systems than in saturated systems because of nonlinear relationships between hydraulic conductivity and water content. These uncertainties in hydraulic conductivity could readily result in order-of-magnitude uncertainties in recharge estimates. Uncertainties in recharge estimates based on tracer data include those associated with measurement of tracer concentrations, estimated inputs of tracers, and assumptions about tracer transport processes. These uncertainties are generally less than those associated with water-budget approaches or methods that use hydraulic-conductivity data.

The tracer techniques such as ^{36}Cl, ^{3}H, $^{3}H/^{3}He$, CFCs, ^{14}C, and Cl can provide integrated, long-term estimates of recharge are. Tracers are very useful for estimating net recharge over long time periods but generally do not provide detail time series information on variations in recharge.

3.9.2. Optimization among Different Factors and Estimating Recharge

One of the difficulties in selecting appropriate technique for quantifying recharge in an area may be that many techniques are restricted to measuring recharge rates within a certain range, which may not be known a priori before the recharge study is undertaken. Therefore, only different approaches that are likely to provide the quantitative estimates of recharge can be considered.

Background information on climate, geomorphology including topography, vegetation, irrigation, soil type and permeability, physiographic setting, subsurface geology and hydrology (water-table depth, gaining vs. losing streams) can be used to develop a conceptual model of recharge at a site and to delineate recharge sources. Then, based on the estimated recharge source and rate, allocated time and money, and considering the limitations/suitability of different approaches and methods, an appropriate method should be selected.

The process of recharge estimation is iterative and involves continual refinement of recharge rates as additional data are collected. Multiple techniques may be used to estimate recharge because of uncertainties associated with each approach.

3.10. Developing a Conceptual Model of Recharge/Conceptualizing a Recharge Model

The fundamental aspect of developing a conceptual model of recharge is understanding the sources of recharge and the spatial and temporal variability in recharge. At first, potential sources of recharge of the target area (or basin) should be identified. The probable sources of recharge for an aquifer include precipitation, surface water (rivers, streams, ponds, playas, lakes, floods), seasonal irrigated area, return flow from irrigation, lateral groundwater flow into the aquifer from outside areas, and cross-formational flow from adjacent aquifers.

The second important aspects are understanding the climate factor (specially rainfall and evapo-transpiration) and hydro-geologic factor. Recharge generally increases with increased precipitation. The seasonal distribution in precipitation may be more important than the average annual precipitation. Winter precipitation is more effective in recharging groundwater than summer precipitation, because if potential evaporation is much greater than precipitation, there should be no groundwater recharge. However, the time scale of the calculations is important. Use of long time scales, such as yearly or monthly, can lead to an underestimation of recharge. Water-budget estimates should be conducted using data and time steps no larger than daily because precipitation at such smaller time scales can greatly exceed evapo-transpiration and result in effective recharge. In addition to climatic factors, recharge is affected by soil texture/structure, and hydraulic conductivity (coarse-grained soils generally result in higher recharge rates than fine-grained soils), land cover (crop-lands generally result in higher recharge than grasslands and shrub-lands), and land use (irrigated lands result in higher recharge than dry-lands).

Geologic formation, up to the aquifer, is important in controlling recharge rate. Considering all the factors and limiting conditions of the factors, a conceptual model of recharge can be developed.

3.11. Challenges in Predictive Relations and Recharge Generalization

Although numerous studies to estimate recharge in specific areas have been carried out worldwide, no systematic attempt has been made to develop generic, predictive relationships for quantifying recharge based on the aforementioned controlling environmental and other factors. This is important for groundwater management and protection. The capabilities of Geographic Information Systems (GIS) can be used to go ahead in this aspect.

An ideal/good recharge model should fulfill the following essential elements/criteria:

- 'Conceptual model' or 'Recharge processes' – Should have an appropriate conceptual model, quantitatively account for detailed hydro-geological processes.
- Accounting all water – Should explicitly account for the water that does not become recharge.
- Sensitiveness' of the uncommon/difficult parameters – The model will not be sensitive to parameters which are difficult to estimate accurately.
- Easiness of use – The model should be easy to operate in terms of data requirement and skill requirement.

Various factors affect the complex groundwater recharge processes, such as rainfall intensity, rainfall frequency, atmospheric water demand (ET), soil type, seasonal vegetation coverage change, irrigation, topography, permeability of vadose zone, and the groundwater level. These factors influence groundwater recharge processes in different ways across climates (individually and interactively), which makes the development of generalized recharge model very difficult/challenging. For example, groundwater is generally considered to occur in topographic highs and discharge in topographic lows in humid regions, whereas in arid alluvial-valley regions groundwater recharge is usually focused in topographic lows, such as channels of ephemeral streams. Groundwater recharge modeling faces particular challenges in arid and semi-arid areas. Here, potential evapotranspiration generally exceeds rainfall, and groundwater recharge depends on high intensity rainfall events, the accumulation of rain-water in depression and streams, and the ability of rain-water to escape evapotranspiration by rapid infiltration through cracks and fissures. Therefore, direct recharge is likely to be less important than localized and indirect recharge, in total aquifer replenishment. Annual rainfall totals may not be good predictor of annual potential recharge, because groundwater recharge results from only infrequent large rainfall events. A process-based model will require extensive data. Data availability is a major issue in developing a continental-scale groundwater recharge model.

3.12. Geological Mapping of the Recharge Areas

The protection of water resources is important for human being with the quality and availability of clean drinking water. Therefore, it is important to identify those parts of the land-surface where our ground water is most likely to be replenished so that we can attempt to protect these vital resources from pollution and any land-use practices that will decrease the quality and availability of clean water.

Groundwater recharge potential maps show land areas characterized by the water-transmitting capabilities of the first 6~10 m below land surface. A groundwater recharge map should be able to differentiate recharge areas on their abilities to contribute water to the aquifer and also provide a means to identify an area's potential impact on adjacent components in the hydrologic system as well as the aquifer. Potential recharge- and discharge-area mapping can be done with GIS. The maps will show how locations of potential recharge and discharge are controlled by a combination of geologic materials and topographic variables.

Groundwater recharge potential maps support decision-making and policy development in land use, water-resources management, wastewater disposal systems development, and environmental permitting in state, county, and local governments. Recently in many countries/states, state law requires that counties and towns with more than certain residents should provide protection to areas with excellent recharge potential in comprehensive land use plans.

Groundwater recharge potential mapping can be done using geologic mapping techniques and with numerous subsurface observations in test borings, wells, borrow pits, natural exposures, and ditches. Hydraulic testing of wells will show the recharge potential categories (excellent, good, fair, poor), which can be used as predictors of the relative amounts and rates at which recharge will occur. The resulting classified potential recharge/discharge area raster data layers can then be mapped on a geo-referenced base for interpretation. To determine the best means for applying in mapping technique to delineating and rating recharge areas, a variety of combinations of layer thickness, number of layers, numbers and types of lithologic and map units, and recharge rating categories can be tested. The lithologic categories should be based on the expected permeabilities of the rock types found in the area and can be derived from interpretations of the geology of the area. The rating criteria should be chosen based on the spatial and/or vertical difference among different data sets (e.g., surface category, layer permeability, etc.) (Figure 3.4).

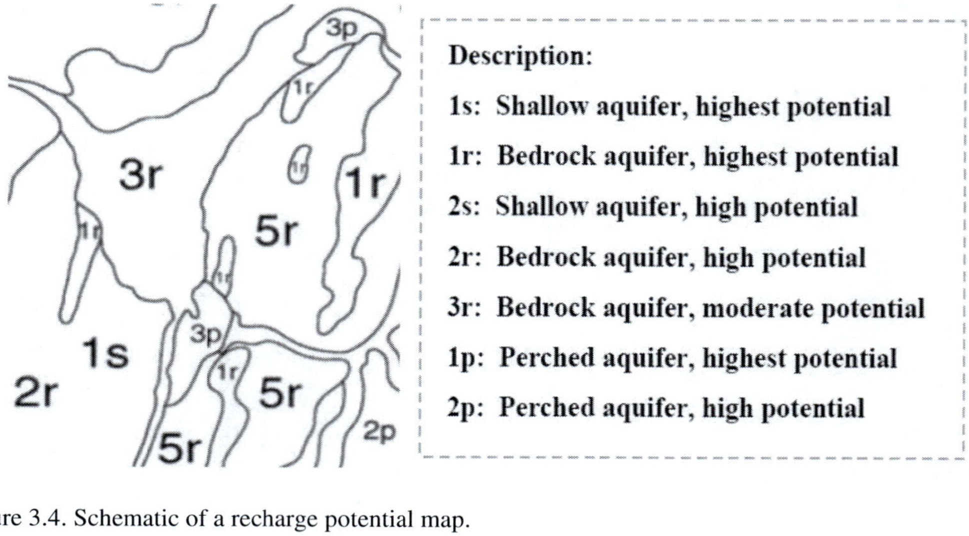

Figure 3.4. Schematic of a recharge potential map.

General Guidelines for Mapping Recharge Area

- Collect rainfall data from climate-monitoring stations, insert in digital map and do averaging/isohyets (using appropriate software, e.g., GIS tool)
- Develop map showing how the land surface is currently in use (residential, agricultural, commercial, wooded, pavement, etc.)
- Develop map showing what kind of soils occur at the land surface, and the extent of wetlands (streams, rivers, lakes, marshes, and bogs).
- Collect geological information through bore log, observation well, etc. and insert in digital map
- Perform rating of recharge (through determining recharge by any of the available suitable method, or from the above information)
- Over-lap the maps (using software capability), set criteria for classifying the infiltration/recharge capabilities
- Produce the output (combined map)
- Delineate the potential recharge area

Detailed three-dimensional geologic maps can be developed and discretized for use as input to a fully three-dimensional groundwater model for variably saturated porous media. The top of the geologic model may be the surface digital elevation model, which can also be used to derive the drainage network that is an important boundary condition in the groundwater flow model. The bottom of the geologic model may the bedrock surface.

Numerical models can be used for the new areas for simulating recharge rates. Because of the association of recharge potential map categories with hydraulic properties, map categories are indicators of how fast contaminants will move and how much water may become contaminated. Numerical models of contaminant transport may be used for different recharge-potential conditions to predict what amount (masses) of contaminants will move more quickly and affect greater volumes of water under higher recharge potential conditions than under lower recharge potential conditions. This information can be used to help prioritize and classify sites for appropriate remedial action.

A practical index method may be developed to produce a map of groundwater availability during drought, indicating the degree of groundwater availability under different rainfall and aquifer conditions, in order to guide groundwater resources management.

3.13. Methods for Estimating/Measuring Components of Water Budget Equation

3.13.1. Evapotranspiration

In a crop field, evaporation from the soil and transpiration through plant occurs simultaneously. It is not easy to measure them separately and for that reason, they are named together. Evaporation from the soil surface of the crop field and the transpiration through the crop plant combinedly is termed as evapotranspiration (ET). The water loss by evapotranspiration from a crop field is influenced by Crop, weather, soil, and management factors.

Direct Measurement of ET by Lysimeter

The evapotranspiration of a crop can be measured through weighing lysimeter, from the difference in weights between two time periods. Periodic thorough water application to the surface of the lysimeter to produce drainage is preferable. Water balance and ET values can be calculated for periods between successive drainage occurrences. With the weighing type lysimeter, hourly or daily estimates of ET can be made.

For non-weighing lysimeter, water application to the lysimeter surface (and to the surrounding buffer area) should be provided whenever the soil moisture depletion or tension of the top soil layers reaches a certain level. Although daily observations of drainage, soil water content, etc. can be made; water balance and evapotranspiration values are normally calculated on a weekly or 10-days basis. The ET for a particular time period can be expressed in the following form of equation:

$$ET_{\Delta T} = SM_{T1} - SM_{T2} + IR + R_e - D - SR \qquad (3.35)$$

Where,

$ET_{\Delta T}$ = The ET for the time period ΔT (between time of SM_1 and SM_2 reading), mm

SM_{T1} = Total soil-water content of the lysimeter soil at the beginning of the time period ΔT, mm

SM_{T2} = Total soil-water content of the lysimeter soil at the end of the time period ΔT, mm

IR = irrigation amount applied within the time period, mm

R_e = Effective rainfall within this time period, mm

D = Deep drainage or percolation within the period, mm

SR = Surface runoff within the period, mm

Total soil moisture within the lysimeter soil column at a particular time may be calculated as:

$$SM_t = \sum_{i=1}^{n} (V_{wi} \times Z_i)/100 \qquad (3.36)$$

where, SM_t = Total profile soil moisture within the lysimeter, in m depth

V_{vi} = Volumetric moisture content in percent for the layer *i*

Z = Depth of soil layer *i*, m

n = Total number of soil layer within the soil column, nos

For gravimetric moisture determination, it took the form:

$$SM_t = \sum_{i=1}^{n} (V_{wi} \times \rho_{si} \times z_i) \times 100 \qquad (3.37)$$

where, V_{wi} = Percent moisture content in weight basis (w/w) for the layer *i*

ρ_{si} = Bulk density of soil of layer *i*

Density of soil-water is considered as 1 gm/cc or 1000 kg/m^3.

Indirect Method

From Field Plot

ET can also be derived from field plot (specially designed, experimental plot) data using water balance equation at the plot level.

From Crop Coefficient

The ET of a crop (ET_c) is computed as follows:

$$ET_c = K_c \times ET_0 \tag{3.38}$$

where,

ET_c = evapotranspiration of a particular crop for a particular period (mm)

K_c = crop coefficient of that crop for the crop period (growth stage) concerned

ET_0 = reference evapotranspiration (mm)

Adjusting for soil moisture, the ET_c becomes:

$$ET_c = K_c \times K_s \times ET_0 \tag{3.39}$$

Where, K_s is a coefficient depended upon available soil moisture.

The value of crop coefficient (K_c) may be obtained from the literature, or may be determined independently from field experimental data of evapotranspiration. The reference evapo-transpiration may be computed from local weather data using various methods (Smith et al., 1992) and tools. More details can be found in Ali (2010b).

3.13.2. Surface Runoff

Surface runoff of a hydrologic basin depends on rainfall amount (intensity & duration), surface vegetation, soil type, land slope, surface depressions, ponds and other natural water storage, channel density, etc. Surface runoff can be measured at the channel inlet by flume, weir, or other measuring structures. Rainfall amount should be averaged over the area.

From Crop Fields

For crop fields having high levees, the surface runoff can be calculated as:

Surface runoff = Total rainfall – (Height of levee – height of water in the field before rainfall)

SCS Runoff Method

The Soil Conservation Service (SCS) [recent 'Natural Resources Conservation Service'] method is an empirical equation. The SCS runoff equation is:

$$Q = \frac{(P - I_a)^2}{(P - I_a) + S} \tag{3.40}$$

Where,
Q = runoff (mm)
P = rainfall (mm)
I_a = initial abstraction (mm)
S = potential maximum retention after runoff begins (mm)
The $I_a$is approximated as (USDA-SCS, 1985):
$I_a = 0.2S$

Thus,

$$Q = \frac{(P - 0.2S)^2}{P + 0.8S} \tag{3.41}$$

The potential retention (*S*) can range from zero on smooth, impervious surface to infinity in deep gravel. For convenience, the 'S-values' were converted to runoff curve numbers (*CN*) by the following transformation.

$$CN = \frac{1000}{10 + S} \tag{3.42}$$

Or,

$$S = \frac{1000}{CN} - 10$$

The curve number is based on the area's hydrologic soil group, land use, treatment and hydrologic condition. The curve number for most US watersheds can be found in Chapter-9 of NEH-4 (National Engineering Handbook, section 4) (USDA-SCS, 1985; USDA-SCS, 1993). The amount of runoff can be found if the rainfall amount and curve number is known.

Peak Runoff from Single Storm Event

The rational method of estimating peak flow on small watershed (< 12 km^2) is based on the assumption that for storms of uniform intensity, distributed evenly over the basin, the maximum rate of runoff occurs when the entire basin area is contributing at the outlet and this rate of runoff is equal to a percentage of the rainfall intensity.

Runoff from an area from a storm event can be calculated using Rational method as:

$$Q = \frac{CIA}{360} \tag{3.43}$$

Where,
Q = runoff (m^3/s)
C = runoff coefficient (dimensionless), representing the ratio of peak runoff rate to average rainfall intensity
I = rainfall intensity (mm/hr)
A = area contributing runoff to the point of consideration (ha)

360 = factor from unit conversion

The runoff coefficient varies with the surface condition. For given storm events of same rainfall intensity, non-constant coefficients of runoff are obtained. The variation is due to the different antecedent moisture conditions. For loose soil (high infiltration) its value is low (0.3 – 0.6) whereas for hard surface its value may be up to 0.8.

3.14. WORKED OUT PROBLEMS

Example 3.1

In a small irrigation farm, the following hydrological data are available for a particular time period:

Rainfall: 350 mm
Surface runoff: 150 mm
Initial ponded water depth = 0 mm
Ponded water depth at the end of time period: 10 mm
Evapotranspiration = 80 mm
Change of moisture storage (increase) within the top 2 meter soil layer = 20 mm
Calculate the deep drainage beyond 2 meter depth.

Solution

We get from water balance eqn.,

Inflow – outflow = Change in storage (i.e., surface + soil column)

Here, inflow is only rainfall; and outflow consists of surface runoff, ET, and deep drainage

i.e., Rainfall– (surface runoff + ET+ Deep drainage) = (Final ponded depth - Initial ponded depth) + ΔSM

or, (350 + 0) – (150 + 80 + Deep drainage) = (10 – 0) + 20

or, 350 – 230 - Deep drainage = 30

or, 350 – 230 -30 = Deep drainage

i.e., Deep drainage = 90 mm (**Ans**).

Example 3.2

In a hydrologic basin, the empirical equation for recharge is expressed as follows:

$$R = -20 + 0.4P$$

Where, R is the recharge (in mm), P is the precipitation (mm).

For a rainfall event, the records of rainfall at difference rainfall stations within the basin are as follows:

	Gauge-1	Gauge-2	Gauge-3	Gauge-4
Rainfall amount (mm)	50	75	80	85
Area under rain-gauge (ha)	100	80	100	75

What is the recharge in the basin from the above rainfall?

Solution

Total area (A) under the basin = 100 + 80 + 100 + 75 = 355 ha

Weighted average rainfall ($P_{w\text{-}av}$) over the basin is:

$$P_{w\text{-}av} = \frac{1}{A}\sum A_i P_i =$$

$$\text{Recharge, } R = -20 + 0.4\, P_{w\text{-}av}$$

$$= -20 + (0.4 \times 71.48)$$

$$= 8.5915 \text{ mm} = 0.0859 \text{ m}$$

Recharge volume = Area × depth = (355 ×10000 m^2) × 0.0859 m = 305000 m^3 (**Ans.**)

Example 3.3

A lysimeter having surface area of 10 m^2 and depth of soil column 2 m, had an initial average moisture content of 30% by volume. The rainfall amount and the ET during the study period were 500 mm and 200 mm, respectively. The average final soil moisture content was 40% by volume. Calculate the amount of recharge.

Solution

Give:

Surface area, A = 10 m2

Depth of soil column, d = 2 m

P = 500 mm

ET = 200 mm

SMi = 30% by vol. (i.e., 30 cm/m depth)

Initial total soil-water, SM_{iT} = 30 (cm/m) × 2 (m) = 60 cm = 600 mm

Final total soil-water, SM_{fT} = 40 (cm/m) × 2 (m) = 80 cm = 800 mm

From water balance equation, recharge,

$R = P + SM_{iT} - SM_{fT} - ET = 500 + 600 - 800 - 200 = 100$ mm (**Ans.**)

Example 3.4

Tritium tracer (^3H) was applied 30 cm below the surface in a grassland before the onset of rainy season. After the rainy season (8 months period), the sampling was done and analyzed for tritium activity. The peak activity was found at 90 cm below the surface. Average soil moisture content between injection point and the peak activity was 30% by volume. The total seasonal rainfall measured at the nearby weather station was 1500 mm.

(a) Determine recharge rate.
(b) Total recharge for the study period.
(c) Assuming negligible rainfall during the rest of the period of the year, estimate the yearly total recharge.
(d) What is the percentage of recharge out of total rainfall?

Solution

Given:
z1= 30 cm
z2 = 90 ,,
Thus, Δz = 90-30 = 60 cm
θ = 30% = 0.3
Δt = 8 months = 240 days
P = 1500 mm = 150 cm

(a) Recharge rate, R = $(\Delta z/\Delta t).\ \theta = (60/240) \times 0.3 = 0.075$ cm/d
(b) Total seasonal recharge = Recharge rate × time period = 0.075 cm/d × 240 d = 18 cm
(c) Recharge, as a percentage of total rainfall = (Recharge/Rainfall) ×100 = (18/150) ×100 = 12%

Example 3.5

A slope gauge having a scale with a slope of 1 vertical to 5 horizontal was installed in a paddy field to measure the water loss in a field. The scale reading at 7:00 a.m. was 75 mm and that after 48 hr was 20 mm. The evapotranspiration and rainfall during this period were 10 mm and 8 mm, respectively. Assuming no horizontal flow (i.e., seepage), calculate the percolation (downward flow) loss of water in the field.

Solution

Difference in scale reading = 75 – 20 = 55 mm
Scale is 1 vertical 5 horizontal
Thus, vertical drop of water level = 55/5 = 11 mm
From water balance equation,

$$\text{Inflow} = \text{outflow} \pm \text{change in storage}$$

i.e., Rainfall = evapotranspiration + percolation ± change in water depth

$$8 = 10 + DP - 11 \text{ [-ve sign for drop, i.e., decrease in storage]}$$

Thus, DP = 8 + 11 – 10 = 9 mm (Ans.)

Example 3.6

A groundwater formation underlying a hydrological basin has a porosity of 50% and specific retention of 20%. During a calendar year, the fluctuation of water level (rise) in an observation well was found 1.5 meters. Determine the amount of deep percolation during the year.

Solution

Specific yield, S_y = porosity – specific retention
= 50% - 20%
= 30%
Deep percolation, $D = h \times S_y$

Here, fluctuation of water-table, h = 1.5 m

Thus, $D = 1.5 \text{ m} \times (30/100)$
= 0.45 m (**Ans.**)

Example 3.7

In a rice field, levees are provided to hold the rainwater. Estimate the amount of surface runoff from a rainfall event of 300 mm, if the height of the levee around the field plot is 20 cm and the height of water in the field before rainfall is 10 cm.

Solution

Runoff amount = rainfall – (height of levee – height of water in the field before rainfall)
= 30 – (20 – 10)
= 30 – 10
= 20 cm **(Ans.)**

Relevant Journals

Hydrogeology Journal
Hydrological Processes
Hydrological Sciences Journal
Geological Sciences Journal
Geological Sci. and Technology Information
Engineering Geology
Research Journal of Environmental and Earth Sciences
Soil Science Society of America Journal

Australian Journal of Soil Research
Applied Geochemistry
Air, Soil and Water Research

Questions/Exercise

1. What do you mean by recharge? Briefly discuss the purpose and significance of recharge estimation.
2. Define the following: Actual recharge, potential recharge, artificial recharge, natural recharge, induced recharge, preferential recharge
3. What are the sources of recharge? Briefly discuss the mechanism of recharge.
4. What are the factors affecting recharge? Briefly narrate 3 of them.
5. Discuss the importances of recharge in relation sustainability of groundwater.
6. Write down the functional form of recharge. Narrate the forms of recharge under different limiting conditions.
7. Classify recharge estimation techniques.
8. Discuss the following methods for recharge estimation and narrate their relative merits and demerits: (a) Water budget, (b) Water-table fluctuation, (c) Seepage meter, (d) Soil-water balance, (e) Zero-flux plane
9. Discuss the merits and demerits of (a) chemical and (b) radio-tracer methods over other methods. Sketch the concentration profile of chemical tracer with depth.
10. Discuss the methods for estimating recharge rate from tracer profile.
11. Write down some empirical equations (with definition of the parameters) for estimating recharge.
12. Do you think that multiple techniques are useful in estimating recharge?
13. Discuss the issues of aquifer vulnerability and recharge estimation.
14. What are the factors to be considered in selecting an appropriate method?
15. How you will develop a conceptual model for recharge?
16. Is it possible to develop generalized predictive model for recharge?
17. Write down the importance of mapping of the recharge areas.
18. Following hydrological data were gathered from a crop field for a particular time period:
    Rainfall: 250 mm
    Surface runoff: 50 mm
    Initial ponded water depth = 0 mm
    Ponded water depth at the end of time period: 4 mm
    Evapotranspiration = 40 mm
    Change of moisture storage (increase) within the top 100 cm soil layer = 5 mm
    Calculate the deep drainage beyond 1 meter depth.
19. In a hydrologic basin, the empirical equation for recharge is expressed as follows:

$$R = 0.63\ (P - 15.28)^{0.76}$$

Where, R is the recharge (in mm), P is the precipitation (mm).

For a rainfall event, the records of rainfall at difference rainfall stations within the basin are as follows:

	Gauge-1	Gauge-2	Gauge-3
Rainfall amount (mm)	60	80	90
Area under rain-gauge (ha)	80	100	85

Calculate the recharge in the basin from the above rainfall.

20. A lysimeter having surface area of 6 m^2 and depth of soil column 1.5 m, had an initial average moisture content of 20% by volume. The rainfall amount and the ET during the study period were 400 mm and 100 mm, respectively. The average final soil moisture content was 30% by volume. Calculate the amount of recharge.
21. To estimate the recharge rate in a hydrologic basin, tritium tracer (^3H) was applied 50 cm below the soil surface before the onset of rainy season. After the rainy season (6 months period), the sampling was done and analyzed for tritium activity. The peak activity was found at 80 cm below the surface. Average soil moisture content between injection point and the peak activity was 25% by volume. The total seasonal rainfall measured at the nearby weather station was 1200 mm. Determine the recharge for the study period.
22. A slope gauge having a scale with a slope of 1 vertical to 4 horizontal was installed in a paddy field to measure the water loss in a field. The scale reading at 9:00 a.m. was 60 mm and that after 48 hr was 30 mm. The evapotranspiration and rainfall during this period were 6 mm and 4 mm, respectively. Assuming no horizontal flow (i.e., seepage), calculate the deep drainage rate of water in the field.
23. A groundwater formation underlying a hydrological basin has a porosity of 55% and specific retention of 15%. During a calendar year, the fluctuation of water level (rise) in an observation well was found 2.5 meters. Determine the recharge during the year.

REFERENCES

Ali, M. H. (2010a). Weather – a driving force in determining irrigation demand. In: *Fundamentals of Irrigation and On-farm Water Management*, Vol. *1*., Springer, NY. pp. 31-105.

Ali, M. H. (2010b). Field water balance. In: Fundamentals of Irrigation and On-farm Water Management, Vol. *1*., Springer, NY. pp. 331-372.

Allison, G. B., Gee, G. W. & Tyler, S. W. (1994). Vadose zone techniques for estimating groundwater recharge in arid and semiarid regions. *Soil Science Society of America Journal,58*, 6–14.

Allison, G. B. & Hughes, M. W. (1983). The use of natural tracers as indicators of soil-water movement in a temperate semi-arid region. *J. Hydrol.*, *60*, 157–173.

Anderson, T. W., Freethey, G. W. & Tucci, P. (1992). Geohydrology and water resources of alluvial basins in South-Central Arizona and parts of adjacent States. U.S. Geological Survey, Professional Paper, 1406-B, pp. 77.

Aranyossy, J. F. & Gaye, C. B. (1992). La recherche du pic de tritium thermonucleaire en zone non saturee profonde sous climate semi-aride pour la mesure de la recharge des nappes: premiere application au Sahel. *C R Acad Sci Paris*, *315*(Ser 2), 637–643.

Arnold, J. G., Allen, P. M., Muttiah, R. & Bernhardt, G. (1995). Automated base flow separation and recession analysis techniques. *Ground Water*, *33*, 1010–1018.

Arnold, J. G., Muttiah, R. S., Srinivasan, R. & Allen, P. M. (2000). Regional estimation of base flow and groundwater recharge in the Upper Mississippi River Basin. *J Hydrol*, *227*, 21–40.

Bajjali, W., Clark, I. & Fritz, P. (1997). The artesian thermal groundwaters of northern Jordan: insights into their recharge history and age. *J Hydrol*, *192*, 355–382.

Bauer, H. H. & Mastin, M. C. (1997). Recharge from precipitation in three small glacial till mantled catchments in the Puget Sound Lowland, Washington. *US Geol Surv Water Resour Invest Rep*, 96–4219, *119*.

Belan, R. A. & Matlock, W. G. (1973). Groundwater recharge from a portion of the Santa Catalina Mountains. In: Proc Meeting Hydrology and Water Resources in Arizona and the Southwest, Arizona Section, American Water Resources Association, and the Hydrology Section, *Arizona Academy of Science*, *Tucson*, *Arizona*, *Rep*, *3*, pp. 33–40.

Chaturvedi, R. S. (1973). "A Note on the Investigation of Ground Water Resources in Western Districts of Uttar Pradesh." *Annual Report*, *U. P. Irrigation Research Institute*, 1973, pp. 86-122.

Christophersen, N. & Hooper, R. P. (1992). Multivariate analysis of stream water chemical data: the use of principal components analysis for the end-member mixing problem. *Water Resour Res*, *28*, 99–107.

Cook, P. G. & Bohlke, J. K. (2000). Determining timescales for groundwater flow and solute transport. In: Cook PG, Herczeg AL (eds) Environmental tracers in subsurface hydrology. *Kluwer*, *Dordrecht*, pp. 1–30.

Cook, P. G. & Solomon, D. K. (1997). Recent advances in dating young groundwater: chlorofluorocarbons, 3H/3He and 85Kr. *J Hydrol*, *191*, 245–265.

Dincer, T., Al-Mugrin, A. & Zimmermann, U. (1974). Study of the infiltration and recharge through the sand dunes in arid zones with special reference to stable isotopes and thermonuclear tritium. *J Hydrol*, *23*, 79–109.

Fayer, M. J. (2000). UNSAT-H Version 3.0: unsaturated soil water and heat flow model, theory, user manual, and examples. Rep 13249. Battelle Pacific Northwest Laboratory, Hanford, Washington DOI 10.1007/s10040-001-0176-2.

Flint, A. L., Flint, L. E., Kwicklis, E. M., Fabryka-Martin, J. T. & Bodvarsson, G. S. (2002). Estimating recharge at Yucca Mountain, Nevada, USA: comparison of methods. Hydrogeol J (in press). DOI 10.1007/s10040-001-0169-1.

Hooper, R. P., Christophersen, N. & Peters, N. E. (1990). Modeling of streamwater chemistry as a mixture of soilwater endmembers – an application to the Panola Mountain catchment, Georgia, USA. *J Hydrol*, *116*, 321–343.

Hsieh, P. A., Wingle, W. & Healy, R. W. (2000). VS2DI – a graphical software package for simulating fluid flow and solute or energy transport in variably saturated porous media. *US Geol Surv Water-Resour Invest Rep*, 99–4130, *16*.

Kumar, C. P. & Seethapathi, P. V. (1987-88). "Effect of Additional Surface Irrigation Supply on Ground Water Regime in Upper Ganga Canal Command Area, Part I - Ground Water

Balance." *National Institute of Hydrology, Case Study Report No.* CS-10 (Secret/Restricted), 1987-88.

Kumar, C. P. & Seethapathi, P. V. Assessment of natural ground water recharge in upper Ganga canal command area.

Nathan, R. J. & McMahon, T. A. (1990). Evaluation of automated techniques for base flow and recession analyses. *Water Resour Res*, *26*, 1465–1473.

Rorabough, M. I. (1964). Estimating changes in bank storage and groundwater contribution to streamflow. *Int Assoc Sci Hydro Publ*, *63*, 432–441.

Ross, P. J. (1990). SWIM – a simulation model for soil water infiltration and movement. Publ 37. CSIRO, Glen Osmond, South Australia.

Sami, K. & Hughes, D. A. (1996). A comparison of recharge estimates to a fractured sedimentary aquifer in South Africa from a chloride mass balance and an integrated surface–subsurface model. *J Hydrol*, *179*, 111–136.

Simmons, C. S. & Meyer, P. D. (2000). A simplified model for the transient water budget of a shallow unsaturated zone. *Water Resour Res*, *36*, 2835–2844.

Simunek, J., Sejna, M. & van Genuchten, M. T. (1996). Hydrus-2D: simulating water flow and solute transport in two-dimensional variably saturated media. International Groundwater Modeling Center, Colorado School of Mines, Golden, Colorado.

Smith, M., Allen, R., Monteith, J. L., Perrier¸ A., Pereira, L. S. & Segeren, A. (1992). Report on the 'Expert Consultation on Revision of FAO Methodologies for Crop Water Requirements.' Food and Agriculture Organization of the United Nations (Land and Water Development Division), Rome, 60 pp.

Sophocleous, M. & Perkins, S. P. (2000). Methodology and application of combined watershed and ground-water models in Kansas. *J Hydrol.*, *236*, 185–201.

Sophocleous, M. A. (1991). "Combining the soil water balance and water-level fluctuation methods to estimate natural ground water recharge - practical aspects." *Journal of hydrology*, Vol. *124*, pp. 229-241.

Theis, C. V. (1937). Amount of groundwater recharge in the Southern High Plains. In: *Proc American Geophysical Union Trans 18th Annu Meeting*, pp. 564–568.

USDA-SCS, (1985). National Engineering handbook, Section 4 – hydrology. Washington, D.C., USDA-SCS.

USDA-SCS, (1993). National Engineering Handbook Series, Part 630, Chapter 4, Storm Rainfall Depth. Washington, D.C., USDA-SCS.

Chapter 4

WATER-WELL CONSTRUCTION AND WELL HYDRAULICS

ABSTRACT

Depending on the permeability of the rocks containing the groundwater, and the pressure under which groundwater is held, the ease of groundwater extraction can vary significantly. Both water quality and aquifer yield determine whether groundwater is appropriate for human consumption, stock water supplies, irrigation, or mining uses. How aquifers respond when water is withdrawn from a well is an important topic in groundwater hydrology. It explains how a well gets its water, how it can deplete adjacent wells, or how it can induce contamination.

4.1. CONSTRUCTION OF WATER-WELL

Wells are used to extract water from aquifers. Basically, a well is a hole drilled into an aquifer. A well consists of a bottom sump, well screen, and well casing (pipe) surrounded by a gravel pack and appropriate surface and borehole seals (**Figure 4.1**). Water enters the well through perforations or openings in the well screen. A pipe and a pump are used to pull water out of the ground, and screen filters out unwanted particles that could clog the pipe. Well design and construction details are determined after a test hole has been completed and the geological zones have been logged.

4.1.1. Importance of Proper Design and Construction of Well

Well design and construction choices affect how efficiently the water in the aquifer enters into the well. The less efficiently water enters into the well, the greater the pumping lift and yearly energy bill, regardless of the pumping plant design and level of pumping plant efficiency. Efficient wells result in less total pumping lift and higher specific well capacity (gpm/ft drawdown). This translates to reduced horsepower needed to lift the water and to decreased hours of pump operation to irrigate a given acreage of crops.

The actual cost of a well depends on the depth to groundwater, the desired well capacity, and choices among a variety of well drilling, well design, well construction, and well development considerations. Choosing the most appropriate approach when investing in a well is site specific and largely dependent upon how extensively the well will be used. Sometimes a simple, less expensive well will suffice and other times added investment in drilling, design, construction, and development will pay back substantially.

4.1.2. Types of Well

Wells come in different shapes and sizes, depending on the type of material the well is drilled into and how much water is being pumped out. There are two main types of wells, each distinguished by the diameter of the bore hole. The two types are bored wells and drilled wells. Bored or shallow wells are usually bored into an unconfined water source, generally found at depths of 30meter or less. Drilled wells may be of two categories, based on the underlying formation:

Consolidated or rock wells: drilled into a formation consisting entirely of a natural rock formation that contains no soil and does not collapse. Their average depth is about 80 m.

Unconsolidated or sand wells: drilled into a formation consisting of soil, sand, gravel or clay material that collapses upon itself.

Bored Wells

A bored well is constructed when low yielding groundwater sources are found relatively close to the surface, usually under 30 m. Bored wells are constructed using a rotary bucket auger. They are usually completed by perforating the casing or using a sand screen with continuous slot openings (**Figure 4.1**).

One advantage of bored wells is the large diameter of the casing, from 45-90 cm (18-36 in.). It provides a water storage reservoir for use during peak demand periods. A disadvantage of utilizing a shallow groundwater aquifer is that it relies on annual precipitation for recharge. Water shortages may occur following long dry periods in summer and extended freeze up during winter months.

Drilled Wells

Drilled wells are smaller in diameter, usually ranging from 15-30 cm, and completed to much greater depths than bored wells, up to several hundred metres. The producing aquifer is generally less susceptible to pollution from surface sources because of the depth. Also, the water supply tends to be more reliable since it is less affected by seasonal weather patterns.

4.1.3. Well Construction

Proper design, construction, development, and completion of the well will result in a long life for the well (as long as half a century or more) and efficient well operation. For many large production wells, a test hole is drilled before well drilling to obtain more detailed information about the depth of water-producing zones, confining beds, well production capabilities, water levels, and groundwater quality.

4.1.3.1. Principal Activities in Well Construction

Well construction involves appropriate site selection, drilling, aquifer identification and yield-test, screen and envelop (gravel pack) design, case and screen setting, sealing (to prevent water from leaking between uncontaminated and contaminated aquifers or from the land surface into the well), well development, installation of pump and power source, and sanitary seal of the well head. Implementation of an aquifer test is normally done for production (large capacity) well.

Site Selection

In constructing a new well, we have to think carefully about the best location, that is, a high point of land with good access and separation from potential contaminants. Wells must be located a safe distance from potential sources of contamination such as fuel storage tanks, stockpiles of chemicals like pesticides or road salt, septic systems, gardens, manure piles, livestock, and roads and driveways.

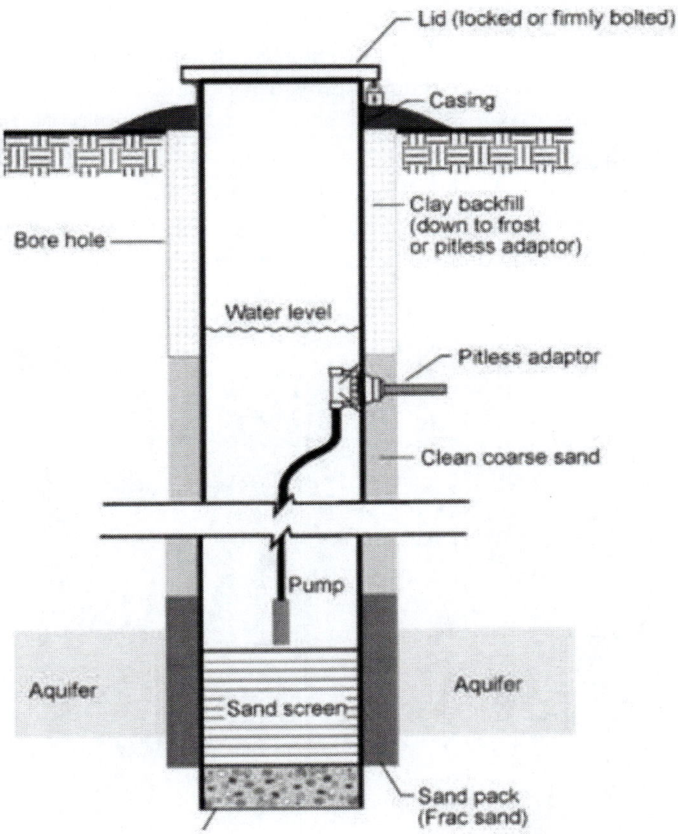

Figure 4.1. Schematic of a bored well.

Drilling

The drilling process itself may last from a few hours (for a shallow, small-diameter well) to several weeks (for a deep, large-diameter well). Sometimes, particularly for large production wells and where water quality is particularly important, the driller will drill a small-diameter pilot hole before drilling the well bore. From information obtained from the pilot hole, a driller or consultant can determine aquifer formations and groundwater quality at various depths and then optimize the final well design for the specific hydrogeological conditions at the site. Appropriate materials (screen, casing, gravel) can then be ordered in a timely fashion prior to the final drilling.

4.1.3.2. Drilling Methods

The most common drilling techniques are auger drilling, rotary, reverse rotary, air rotary, and cable tool.

So, there are two primary methods of drilling:

- Rotary
- Cable tool.

In unconsolidated and semi-consolidated materials, (reverse) rotary and cable tool methods are most commonly employed. Hardrock wells generally are drilled with air rotary drilling equipment. Auger drilling is often employed for shallow wells that are not used as supply well.

Rotary drilled wells are constructed using a drill bit on the end of a rotating drill-stem. Drilling fluid or air is circulated down through the drillstem in the hole and back to the surface to remove cuttings (**Figure 4.2**). Rotary drilling rigs operate quickly and can reach depths of over 300 m, with casing diameters of 10 - 45 cm.

Figure 4.2. View of rotary drilling.

Cable tool drilled wells are constructed by lifting and dropping a heavy drill bit in the bore hole. The resulting loose material, mixed with water, is removed using a bailer or sand pump. This method, also called percussion drilling, reaches depths up to 300 m. Well diameters can range from 10 to 45 cm. The drilling rate is typically much slower than for a rotary rig, but when aquifers are low yielding, they may be more easily identified using this method.

During drilling, drillers must keep a detailed log of the drill cuttings obtained from the advancing borehole. In addition, after the drilling has been completed but before the well is installed, it is often desirable to obtain more detailed data on the subsurface geology by taking geophysical measurements in the borehole.

4.1.3.3. Definition of Relevant Terminologies

Casing: Steel or plastic pipe placed in the borehole to keep it from collapsing. The casing is sealed to the borehole wall near the land surface with the annular seal.

Screen or perforations: All wells are open to the aquifer so that water can enter the well. Well completions vary from "*open hole*" in consolidated rock that does not need a casing, to "*open bottom*" where the only way for the water to enter the well is through the end of the casing. However, many wells have some sort of well screen installed or perforations cut into the casing through which water can enter. The openings must be correctly sized so that water will enter, but sand and other aquifer materials do not.

Total depth: The total depth of the well is the distance from land surface to the bottom.

Tailpipe and end cap: Wells that are completed with well screens may have a tailpipe installed below the screen. The tailpipe provides a place where sand that may enter the well through the screen can settle away from the pump. The end cap forces all water to enter the well through the well screen. Most wells that are completed with perforations may not have a tailpipe.

4.2. WELL DESIGN

Designing a water well involves selecting the proper dimensional factors and the proper materials to insure the optimum combination of performance, long service life, and reasonable cost. The overall objective of the design is to create a structurally stable, long-lasting, efficient well that has enough space to house pumps or other extraction devices, allows groundwater to move effortlessly and sediment-free from the aquifer into the well at the desired volume and quality, and prevents bacterial growth and material decay in the well.

4.2.1. Design Elements and Design Considerations

There are many components to well design which must be taken into account. Decisions should be made about:

- Well depth
- Type of well

- Casing material, size and wall thickness
- Intake (screen) design
- Filter material design
- Formation seal
- Monitoring and preventive maintenance provisions.

Well Depth

During the test hole drilling, the drilling contractor will complete a formation log. Soil and rock samples are taken at various depths and the type of geologic material is recorded. This allows to identify aquifers with the best potential for water supply. Sometimes it is also run an electric or gamma-ray log in the test hole to further define the geology. This gives more accurate information about aquifer location.

Generally a well is completed to the bottom of the aquifer. This allows more of the aquifer to be utilized and ensures the highest possible production from the well. There are two reasons for this. First, more of the aquifer thickness can be used as the intake portion of the well resulting in higher specific capacity. Second, more drawdown can be made available, permitting a greater yield.

One departure from casing to the bottom of the aquifer is when the bottom of the aquifer contains fine sand, or is known to contain poor quality water. In this case the well should be completed above the fine materials, or to a depth which will avoid poor quality water.

Casing Size and Material Type

Choosing the right diameter for the well casing and screen is important because it has an effect on both the cost of the well and its efficiency. Decisions about the diameter and type of well casing are made after considering the following:

- Aquifer characteristics
- Hydraulic factors that influence well performance
- Drilling method
- Well depth
- Cost (in discussion with the well owner)
- Pump diameter (The casing must be large enough to house the pump and allow sufficient clearance for installation and efficient operation).

For deep wells and wells larger than 4 inches, the diameter is usually chosen to accommodate the size of pump required. The diameter should be 2 nominal sizes larger than the pump to allow sufficient clearance for ease of installation and operation. If the screened interval is to be at much lower depths than the pump, the casing below the pump may be of a smaller diameter. For wells smaller than 4 inches it is satisfactory to select casing which is only 1 size larger than the pump.

In shallow wells where the pump is directly connected to the top of the well casing or connected to a suction pipe inside the well, the casing diameter should be selected in relation to the diameter of the suction inlet of the pump.

If a submersible pump is going to be used, the casing must have an inside diameter of at least 10.16 cm (4 in.), by law. It is recommended that the casing be at least one nominal size larger than the outside diameter of the pump. The more space there is between the pump and the casing, the easier it will be to service and repair the pump in the future.

There are two common materials used for casing: steel and plastic (PVC). Steel casing is the strongest but is susceptible to corrosion. Plastic casing is becoming more popular because of its resistance to corrosion. For PVC, maximum depths are usually less than 200 feet and, for strength, the diameters are not larger than 6 inches. All casing must be new and uncontaminated. Plastic casing must be made of virgin resin, not recycled material.

Well Screen

The greatest influence on the efficient performance of a well is the design and construction of the well screen. The well screen serves as the intake section of the well allowing water and is considered adequate only when it allows sand-free water to flow into the well with a minimum loss of head. A properly designed well screen combines the highest percentage of open area with adequate strength. Water moves from the aquifer into the well through either a manufactured screen or mechanically slotted or perforated casing.

Screens are manufactured with regularly shaped and sized openings. They are engineered to allow the maximum amount of water in with minimal entry of formation sediments. Stainless steel screens are the most widely used because they are strong and relatively able to withstand corrosive water. Screens are manufactured with various slot sizes and shapes to match the characteristics of the aquifer.

Slotted or perforated casing or liner is made by creating openings using a cutting tool or drill. Pre-slotted plastic pipe is also available. Slot openings and perforations are spaced further apart than screen openings. This reduces the amount of open area to allow water into the well. The openings tend to vary in size and may have rough edges depending on how they were made. This impedes the flow of water into the well and may not hold back the formation sediments.

After examining the cuttings from the borehole, a judgement should be made whether to use a screen, or slotted or perforated casing/liner. While a screen is the more expensive alternative, it is necessary if the aquifer is composed of loose material such as fine sand, gravel or soft sandstone. A slotted or perforated casing/liner can be used when the aquifer formation is more consolidated, such as hard sandstone or fractured shale.

After a choice is made between a screen, or slotted or perforated casing/liner other decisions will be made regarding:

- Size of slot opening
- Total area of screen or perforation that is exposed to the aquifer
- Placement of the screen or perforations within the aquifer.

Slot Size Openings

The slot openings must be small enough to permit easy entry of water into the well while keeping out sediment. The slot size chosen will depend on the particle size of the earth materials in the producing aquifer (also termed as formation materials).

Typically, select a slot size that allows 60 percent of the aquifer material to pass through during the well development phase of drilling. The remaining 40 percent, comprising the coarsest materials, will form a natural filter pack around the perforations or screen.

Screen Length, Pattern, Total Open Area, and Placement

A well screen is an engineered device that is used in wells to help maximize inflow from the aquifer and allow for long-term satisfactory operation of the well. Well screens are typically installed in wells where the aquifer is comprised of loose or unstable material. The screen prevents rock fragments from entering the well, helps support the wall of the well and allows water to enter slowly. Turbulent flow can more easily transport unwanted rock particles and agitated water may release minerals and clog up the well.

The purpose of the screen is to keep sand and gravel from the gravel pack out of the well while providing ample water flow to enter the casing. The screen should also be designed to allow the well to be properly developed. Slotted, louvered, and bridge-slotted screens and continuous wire wrap screens are the most common types. Slotted screens provide poor open area. They are not well suited for proper well development and maintenance, and are therefore not recommended. Wire wrap screens or pipe-based wire wrap screens give the best performance. The additional cost of wire wrap screens can be offset if only screen sections are installed in the most productive formations along the borehole.

Irrigation wells must be capable of producing adequate water during peak seasonal use and under drought conditions. Without a reliable, efficient, and economical supply of water, the entire irrigation system, regardless of the most sophisticated well head equipment design, becomes nearly useless.

The purposes of the blank well casing between and above the well screens are to prevent fine and very fine formation particles from entering the well, to provide an open pathway from the aquifer to the surface, to provide a proper housing for the pump, and to protect the pumped ground water from interaction with shallower ground water that may be of lower quality. In some cases it may be necessary to place additional casing to seal off parts of the drilled hole where, for example, the water may have high iron content, or some other unwanted chemical attribute.

For naturally developed wells, the size(s) of well screen slot openings will depend on the gradation of the sand, and slot openings are selected using the results of sieve analyses of water-bearing formation samples. A sieve analysis curve should be plotted for each sand sample. The size of the screen opening is selected so that the screen will retain 40-50% of the sand.

The total area of the slot openings is dependent on the length and diameter of the screen. While the length of the screen is variable, the diameter of the screen is determined by the diameter of the well casing. The yield from a well increases with an increase in screen diameter but not proportionately so. While high specific capacity is obtained by using as long a screen as possible, short screens provide more available drawdown. These two conflicting aims are satisfied, in part, by using an efficient well screen.

The amount of open area of the screen or slotted or perforated casing/liner must be calculated to ensure the water from the aquifer does not enter the well too quickly. A larger amount of open area allows the water to enter the well at a slower rate, causing a lower drop in pressure in the water as it moves into the well. If the water flows too quickly, there will be problems with incrustation. Incrustation is a buildup that occurs when dissolved minerals in the groundwater come out of solution and deposit on the screen or casing. As a result, the perforations get plugged and water cannot enter the well at the same rate, and the yield from the well will be reduced. The pore spaces in the aquifer immediately adjacent to the perforations may also get plugged with fine material which could result in yield reduction.

The screen or perforations on the casing/liner must be placed adjacent to the aquifer. If improperly placed, the well may produce fine sediment which will plug plumbing fixtures and cause excessive wear on the pump. If geophysical logging equipment is used to accurately identify the boundaries of the aquifer, the exact placement will be easier.

Screen Material

Depending on the results of preliminary investigation, the well screen should be fabricated of materials that are as corrosion resistant as necessary. If the screen corrodes, sand and/or gravel will enter the well, which may eventually require either replacement of screen or drilling a new well.

Corrosion of screens can occur from bimetallic corrosion if two different metals have been used in the fabrication; therefore, bimetallic screen should always be avoided. Water with high total dissolved solids accelerates this type of corrosion because the water is a more effective electrolyte. Corrosion can also occur from dissolved gases in the water such as oxygen, carbon dioxide, and hydrogen sulfide.

The choice of the well screen material is sometimes based on strength requirements regarding column load and collapse pressure. When a long screen supports a considerable weight of pipe, it functions as a slender column. The pressure of the formation and materials caving into the well pipe can squeeze the screen. Therefore, the well material should be able to withstand the pressure. It is impossible to accurately determine or calculate earth pressures with depth but generally greater strength is needed at greater depths.

Filter Material

It is important to match the grain size of the filter material with the size of the slot openings of the screen to attain maximum yield from the well. Gravel packing makes the zone immediately surrounding the well screen more permeable by removing the formation materials and replacing them with artificially graded coarser materials. The size of this artificially graded gravel (filter material) should be chosen so that it retains essentially all of the formation particles. The well screen slot opening size is then selected to retain the gravel pack. Gravel pack design includes specification of gradation, thickness, and quality of the gravel pack material. Part of the aquifer thickness to be screened should be evaluated by examining the samples collected during the test hole drilling.

A sieve analysis should be prepared for the strata comprising the portion of the aquifer where the screen will be set. Results of sieve analysis for the finest stratum should be used to design the gravel pack grading. It is best to design as uniform a pack as possible. A uniform gravel pack has significantly greater permeability and is easier to install without segregation. The gravel pack material should consist of clean and well-rounded grains that are smooth. These characteristics increase the permeability and porosity of the gravel pack.

The gravel placement should be completed in one continuous operation.

Casing Materials

The casing (or well pipe) is a very critical element in well construction. Casing may be metallic (black iron or galvanized steel) or nonmetallic (polyvinyl chloride (PVC) or ABS plastic). It must be adequately seated in a consolidated formation (limestone, sandstone, etc.) or attached to a screen suitably designed and situated in unconsolidated materials (shell, sand, gravel, etc.). The purpose of casing is to seal off materials that may enter the pumping system from strata other than the aquifer selected and prevent mixing between aquifers. To prevent contamination from surface flow into the well, the casing must be extended above surface flood water levels, and the top portion must be grouted with cement or an approved alternative material.

4.2.2. Design Criteria and Procedure

Water wells need to be designed to suit the geologic conditions, the purpose for needing the water and to comply with local regulations. The final design is subject to site-specific observations made in the test hole or during the well drilling.

Diameter of Slot/SCREEN opening

For naturally developed well (no artificial gravel pack):

$$D_s < D_{0.50p}$$

Where, D_s is the slot size, $D_{0.50p}$ is the diameter that retain 50% of the formation materials (sands).

Artificial gravel pack: Typically the slot size of the screen (D_s) is selected so that greater than 85 percent of the artificial pack material remains outside of the screen. i.e.,

$$D_s < D_{0.85p}$$

Where, $D_{0.85p}$ = diameter that retain 85% pack/filter materials

Screen Open Area

Screen open area (A) should be calculated as:

$$A = Q/V$$

Where, Q is the well discharge, and V is the maximum flow velocity of water to screen (0.03 m/s).

Then, the total screen area,

$$A_T = A/(E_a \times P_s)$$

Where, P_s is the percent open area in the screen, E_a is the effective open area considered for design (50%) [as some of the openings may be clogged after operation of some period].

Length of Screen

Length of screen is selected based on other two factors - screen diameter (d) and open area (A), as:

$$L = A_T/(\pi d)$$

Available drawdown is also considered in selecting length. Drawdown during pumping (specially in dry period) must be above the screen.

Position of Screen

Artesian aquifer: Lower 70% – 80% of the water bearing strata.
Watertable aquifer: Bottom 30% - 60% of the aquifer.

Screen Material

Factors to be considered are corrosivity of water, strength requirement (length of pipe and hence load on it), and economy.
For non-corrosive water: plastic (PVC), galvanized steel, stainless steel,
For corrosive water: stainless steel

Hydraulic Criteria/Velocity of Water

$$V_w = Q/A_T$$

Where, V_w is the velocity of water entering the screen, Q is the desired well yield/discharge, and A_T is the effective total open area in the screen.

$$V_w < 0.03 \text{ m/s}$$

Diameter of Screen Pipe, Vertical Velocity

The screen diameter should be such that the velocity of vertical water flow to the pump does not exceeds 1.5 meter per second. That is,

$$V_v < 1.5 \text{ m/s}$$

Where, V_v is the vertical velocity of water within the screen pipe towards the pump.

Gravel Pack/Filter Material

Grading of filter material: Based on screen opening and aquifer materials.
$D_{f(85\%)}$= 4 to 6 times of aquifer materials (D_{aqf})

Uniformity coefficient (D_{40}/D_{90}): < 2.5
Thickness of filter: 0.08 to 0.2 m.

4.3. Well Completion and Development

4.3.1. Well Completion

A properly constructed well forms an effective barrier against surface run-off that may enter and contaminate the well. Water must infiltrate and pass downward through the soil and/or rock before it can reach the aquifer. Over the years, well design has improved to reflect advances in technology and our understanding of potential pathways of contamination.

Well Casing and Sealing

New wells should be lined with a watertight casing designed to prevent the walls of the well from collapsing. Well casings must be of sufficient length to keep contaminants out of the well water. Steel casings are typically used, but casings can also be made from plastic. The Well Drilling regulations in a region (or State), under the Water Resources Act, outline minimum construction requirements for drilled wells.

The Annular Seal

When a well is drilled the hole in the ground is bigger than the well casing. The resulting gap – the annular space – must be filled with a watertight sealant such as bentonite that does not shrink or crack under the ground. For maximum protection, the sealant should extend the full length of the casing. The annular seal serves as a barrier to run-off, surface water, and near-surface waters that could otherwise travel down the outside of the casing and contaminate the aquifer.

Annulus Seal

Sealing the well protects the well's producing zone from contamination. The diameter of the bore hole is usually slightly larger than the casing being installed. The space between the bore hole and the casing is called the annulus of the well. It must be sealed to prevent any surface contamination from migrating downward and contaminating the water supply. It also prevents any mixing of poor quality aquifers with the producing aquifer of the well (**Figure 4.3**). It is filled with gravel or coarse sand and fine sand particles. The uppermost section of the annulus is normally sealed with a bentonite clay and cement grout to ensure that no water or contamination can enter the annulus from the surface. To isolate the producing zone of the well, the annulus is filled from immediately above the perforated zone to the ground surface.

Well Cap

The drilled well must be capped with a commercially manufactured vermin-proof well cap. Vermin-proof caps have rubber gaskets and screened vents inside to prevent entry of "foreign material" such as vermin, insects, and decaying plant material. Loose fitting caps found on older wells make these wells a comfortable home for insects and vermin.

Filter Material

If the aquifer formation does not naturally have any relatively coarse particles to form a filter, it may be necessary to install an artificial filter pack. The pack is placed around the screen or perforations so the well can be developed. For example, this procedure is necessary when the aquifer is composed of fine sand and the individual grains are uniform in size.

Once the well has been drilled and the equipment is in place, there are several procedures the drilling contractor must complete before the well is ready to use. The drilling contractor is responsible for:

- Well development
- Disinfecting the well
- Conducting a yield test.

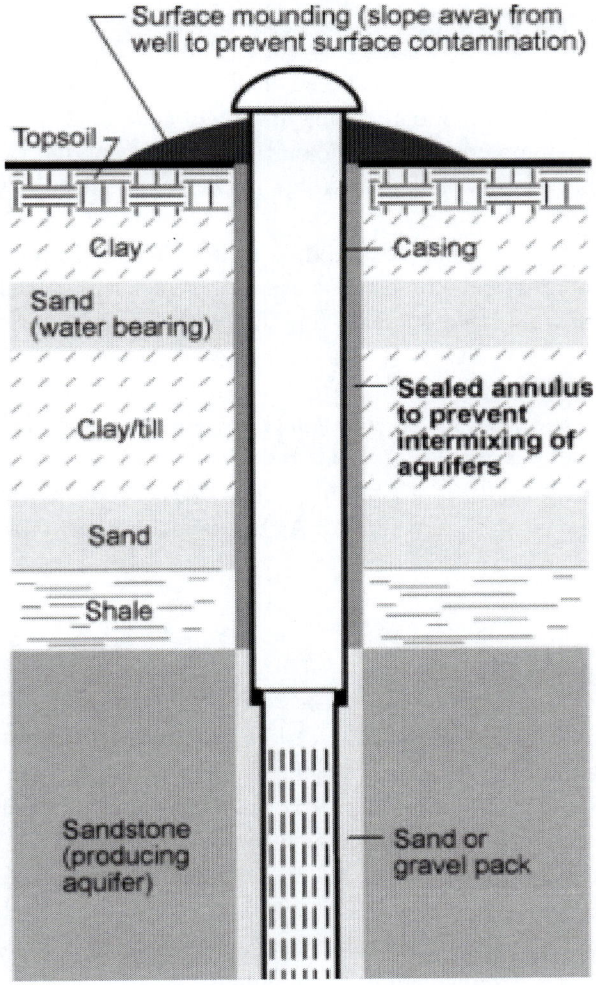

Figure 4.3. Sketch of well seal.

4.3.2. Well Development

After the well screen, well casing, and gravel pack have been installed, the well is *developed* to clean the borehole and casing of drilling fluid and to properly settle the gravel pack around the well screen. A properly developed gravel pack keeps fine sediments out of the well and provides a clean and unrestricted flow path for ground water. Proper well design and good well development will result in lower pumping costs, a longer pump life, and fewer biological problems such as iron-bacteria and slime build-up. Poorly designed and underdeveloped wells are subject to more frequent pump failures because sand and fines enter the well and cause significantly more wear and tear on pump turbines.

Well development is the process of removing fine sediment and drilling fluid from the area immediately surrounding the perforations. This increases the well's ability to produce water and maximize production from the aquifer. Jetting, surging, backwashing, and overpumping (higher than the design pumping rate) are methods used to develop a well. A typical method for well development is to surge *or* jet water or air in- and out- of the well screen openings. This procedure may take several days or perhaps longer, depending on the size and depth of the well. Water or air is surged back and forth through the perforations. Any fine materials that are in the formation become dislodged and are pumped or bailed from the well. This procedure is continued until no fine particles remain and the water is clear. Coarser particles are left behind to form a natural filter pack around the screen, slot openings or perforations.

Testing for Sand. The sand content should be tested after development and performance (pump) testing. Sand production should be measured by a centrifugal sand sampler or other acceptable means. Following development (i.e., stabilization of the formation and/or gravel pack) and pump testing, the sand content should not exceed a concentration of 5 ppm (parts per million) by weight 15 minutes after the start of pumping. Sand production exceeding this limit indicates that the well may not be completely developed or may not have been properly designed. In that event, redevelopment may be appropriate or as an alternative, a sand separator installed. In existing wells should this value be exceeded significantly, rehabilitation (redevelopment) or repair is probably needed. Again, as an alternative, a sand separator may need to be installed.

4.3.3. Disinfection of Well

Some Provincial regulations require the drilling contractor to disinfect new wells with chlorine. The concentration is calculated on the volume of water that is in the well. The concentration must be at least 200 milligrams of chlorine per litre of water present throughout the water in the well and must be left in the well for at least 12 hours to ensure any bacteria present are destroyed. Chlorination is done after the pumping equipment is installed and before the well is put into production. The yield test provides a benchmark of the well's performance. Repeating this test at a later date can be used to assess any changing conditions of the well and determine when maintenance is required.

4.3.4. Economic Considerations

Some of the most important things that should be considered in well construction and design are:

(i) Desired life expectancy of the well
(ii) Appraisal value of the land
(iii) Power cost (power/fuel cost)
(iv) Water quality

In designing a well, considerations should be given concerning the use of top/bottom water. The water below the depth of the well has a much less value when it is not tapped by the original well; whereas, if that same water is tapped by the original well (although possibly not immediately being pumped), has much value in the future. The decision on how much of the available producing aquifer to incorporate into any given well design should be based on expected well life and that affect on land appraisal value. The shallow well can be looked at as a band-aid solution, the deeper wells would be a valuable investment in that particular property.

4.4. WELL HYDRAULICS

Groundwater is particularly important as a water resource in semi-arid to arid parts of the world, and also in humid region, where rainfall is too infrequent or inadequate, or unevenly distributed throughout the year, to reliably meet water needs. Often such groundwater resources have accumulated over long periods and are replenished only when rainfall is sufficient to infiltrate soil and rock. This means groundwater can be a finite, or slowly replenished resource. In more temperate areas where rainfall rates are higher, groundwater may be replenished on a regular basis and extraction can be managed on a renewable basis. However, in many instances groundwater use exceeds the rate at which groundwater is replenished. The yield characteristics of groundwater resources vary considerably across the continent.

When water is withdrawn from a well, its water level drops. When the water level falls below the water level of the surrounding aquifer, ground water flows into the well. The movement of water from an aquifer into a well alters the surface of the aquifer around the well. It forms what is called a cone of depression. A cone of depression is a funnel-shaped drop in the aquifer's surface (**Figure 4.4**). The well itself penetrates the bottom of the cone. Within a cone of depression, all groundwater flows to the well. The outer limits of the cone is termed as the well's area of influence.

4.4.1. Relevant Terminologies

Drawdown: The drawdown in a well is the difference between the pumping water level and the static (non-pumping) water level. Drawdown begins when the pump is turned on, and increases until the well reaches "*steady state*" sometime later. Therefore, drawdown measurements are usually reported along with the amount of time that has elapsed since pumping began. For example, "*The drawdown was 2 m, half-hour after pumping began.*"

Drawdown cone: The depression in the water table near the well that is caused by pumping is called the "*drawdown cone*" or sometimes the "*cone of depression.*" When the well is pumping, water levels are drawn down most near the well and the amount of drawdown decreases as the distance from the well increases. At some distance from the well at any given time there is a point at which the pumping does not change the water table and the drawdown is zero.

Measuring point: Water levels in wells are usually reported as depths below land surface, although the measuring point can be any convenient fixed place near the top of the well. In this drawing the measuring point is the top of the casing. The altitude of the measuring point is commonly recorded so that static water levels can also be reported as altitudes.

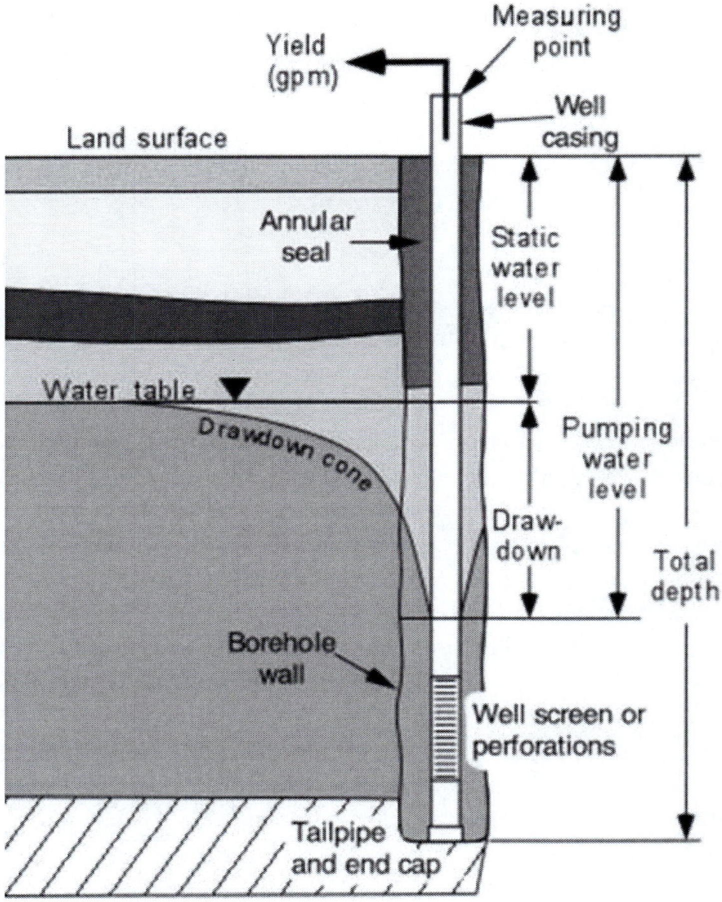

Figure 4.4. Schematic of cone of depression during pumping.

Pumping water level: The pumping water level is the distance from the land surface (or measuring point) to the water in the well while it is pumping. The time that the pumping water level is measured is usually recorded also. For example, "*The pumping water level was 28meter below land surface, 1 hour after pumping began.*"

Static water level: The static water level is the distance from the land surface (or the measuring point) to the water in the well under non-pumping (static) conditions. Static water levels can be influenced by climatic conditions (by evapotranspiration) and pumping of nearby wells and are often measured repeatedly to gain information about how aquifers react to climatic change and development.

Watertable: The top of the saturated part of an unconfined (also known as water-table) aquifer. Below the water-table, pore spaces (or fractures) in the geologic media are filled with water. Above the water-table, the pore spaces are filled with air.

Specific Capacity

The yield of the well per unit of drawdown, usually expressed as gallons per minute (gpm) per foot of drawdown, is called specific capacity. It is obtained by dividing the pumping rate by the drawdown for a specific pumping period. For example, if the pumping rate is 1500 gpm and the drawdown is 20 feet, the specific capacity of the well is 75 gpm per foot of drawdown.

Well Capacity or Yield

The volume of water per unit of time discharged from a well is its capacity. Well capacity is usually measured as the pumping rate in gallons per minute (gpm) or cubic feet per second (cfs).

Well Efficiency

The ratio of the actual specific capacity of a well at the design yield to the maximum specific capacity possible calculated from formation hydraulic characteristics and well geometry is the well efficiency. This is the same as the ratio of the theoretical drawdown to obtain design yield from a 100% efficient well to the actual drawdown measured in the well when producing at the design yield. Efficiency is usually expressed as a percent. The difference (drawdown increase) between the theoretical drawdown and actual drawdown represents the head loss required to force water through the well screen. This head loss should be a minimum. Well efficiency should not be confused with pump efficiency. Pump efficiency is a characteristic of the pump only and is completely independent of well efficiency. For example, the pump efficiency may be 75% while the well efficiency of a poorly designed and constructed well may be only 45%.

Test Hole Assessment – Drilling a test hole provides an opportunity to collect and assess certain design-critical information, such as formation samples for sieve analysis (screen and filter pack design) and lithologic and geophysical data for screen and seal placement.

Well Development Assessment – Properly developing a new water well is one of the most important steps in bringing it online. Development is both a scientific and intuitive process of increasing the specific capacity of the well, while reducing the residual plugging from drilling fluids and sediment in the filter pack.

4.4.2. Well Yield in Aquifer

4.4.2.1. Flow of Water to Well in Unconfined Aquifer

Well discharge equations for equilibrium conditions in unconfined aquifer have been derived by several investigators over the years. These equations relating well discharge to drawdown ignored vertical component of flow.

Theim Equation

Let us consider flow of water to well in unconfined aquifer as shown in **Figure 4.5**. According to Darcy's law:

$$Q = KiA$$

Considering elementary horizontal contributing extent of 'x' and vertical extent of 'y,' the area contributing to flow is 'xy.' Thus,

$$Q = K.\frac{dy}{dx}.(2\pi x.y)$$

Or, $Q.\frac{dx}{x} = 2\pi K.y.dy$

Integrating over the area of screen diameter to 'area of influence' along the x-axis, and from the top of well screen to the free watertable, we obtain:

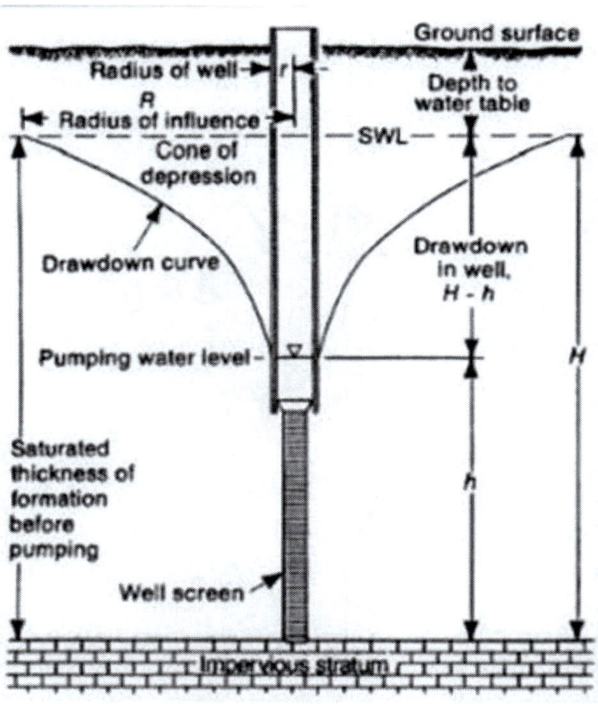

Figure 4.5. Schematic of WT during puming in unconfined aquifer.

$$Q\int_{r1}^{r2}\frac{dx}{x} = 2\pi K\int_{h1}^{h2} y.dy$$

Which reduces to:

$$Q = \frac{\pi K(h_2^2 - h_1^2)}{2.3\ log_{10}\frac{r_2}{r_1}} \tag{4.1}$$

Considering the horizontal extent between well screen *or* face of the well (where drawdown is maximum) and at the upper most influential area (radius of influence), vertical extent from top of the well screen (h_w) to saturated thickness of the aquifer (H) we can write:

$$Q = \frac{\pi K(H^2 - h_w^2)}{2.3\ log_{10}\frac{R}{r_w}} \tag{4.2}$$

Where,

Q = well discharge (m^3/d)

K = hydraulic conductivity of the aquifer ($m^3/m^2/d$)

H = saturated thickness of the aquifer [i.e., height of water-table from the bottom of the screenat steady-state (non-pumping) condition] (m)

h_w= height of water-table at the well-face from the bottom of the screen at pumping condition (m)

R = radius of influence due to pumping (m)

r_w = radius of the screen pipe (m)

This is the well-known Theim equation (Theim, 1906; originally proposed by Dupuit (1863)) for well discharge from unconfined aquifer.

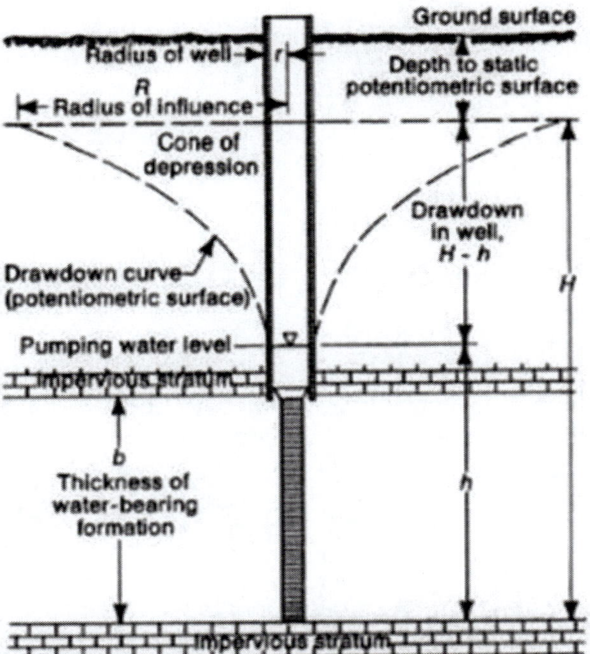

Figure 4.6. Schematic of cone of depression and radius of influence in confined aquifer.

4.4.2.2. Flow of Water to Well in Confined AQUIFER

Theis Equation

Theis (1935) derived equation for discharge from confined aquifer under customary assumptions. Assumptions of Theis equation are as follows:

- Aquifer is confined top and bottom
- There is no source of recharge to aquifer
- The Aquifer is compressible and water is released instantaneously from the aquifer as the head is lowered
- The well is pumping at a constant rate.

The well discharge in a confined aquifer (**Figure 4.6**) is expressed as:

$$Q = \frac{2.73\,T(h-h_w)}{log_{10}\frac{R}{r_w}} \tag{4.3}$$

Where,

Q = well discharge (m^3/d)

T = transmissivity of the aquifer (T = Kb) ($m^3/m/d$)

h = height of peizometricwatertable from the bottom of the screen at steady-state (non-pumping) condition (m)

h_w = height of watertable at the well from the bottom of the screen during stabilized drawdown (m)

R = radius of influence due to pumping (m)

r_w = radius of the screen pipe (m)

Derivation of the Equation

Using Darcy's law:

$$Q = KiA = K.\frac{dy}{dx}.(2\pi x.b)$$

Or, $Q.\frac{dx}{x} = 2\pi Kb.dy$

Integrating over the area of screen diameter to 'area of influence' along the x-axis, and from the top of well screen to the piezometric water-table, we obtain:

$$Q\int_{r_w}^{R}\frac{dx}{x} = 2\pi Kb\int_{h_1}^{h_2} dy$$

Which reduces to:

$$Q = \frac{2.73Kb(h_2-h_1)}{log_{10}\frac{R}{r_w}}$$

4.5. Pumping Test/Well Yield Test and Determination of Aquifer Parameters

The main objective of aquifer test is to determine the ability of the aquifer to transmit and store water (which could be used to design capacity of the pump in new well) and to determine adverse effects on existing wells created by new withdrawals.

4.5.1. Relevant Terminologies

Residual Drawdown

After pumping is stopped, the water level rises and approaches the original static water level. During water-level recovery, the distance between the recovery water level (RWL) and the initial static water level (SWL) is called residual drawdown.

Residual Drawdown = RWL – SWL

Specific Capacity

Specific capacity is the pumping rate divided by the drawdown (L/min/m or GPM/foot). It is a measure of the drawdown caused by the pumping rate and is used as a basis for determining the well's performance. The specific capacity usually changes with both pumping rate and time.

Specific Capacity of a well = Pumping Rate/Drawdown

Well Efficiency

Well efficiency is defined as the ratio of the theoretical drawdown in the formation to the actual drawdown in the well. The difference between the two is caused by frictional energy losses of the water as it moves from within the formation to the pump intake. Thus, well efficiency describes the effectiveness of a well in yielding water. Well efficiency should not be confused with pumping-plant (motor and pump) or "wire-to-water" efficiency used to measure pumping-plant performance.

4.5.2. Perspectives of Pumping Test

Once the well is completed and developed, it is a good practice to conduct an *aquifer test* (or *pump test*). Aquifer tests are used to determine the efficiency and capacity of the well and to provide information about the permeability of the aquifer. The information about the pumping rate and resulting pumping water levels are also critical in determining pump size and optimize well-field production, providing maximum groundwater withdrawal while limiting stress on the aquifer system.

Once the yield test is complete, it can be decided at what rate the aquifer can be pumped without lowering the water level below the top boundary of the aquifer, the top of the perforations or below the pump intake. The pump that is installed in the well should have a capacity equal to, or less than, the rate at which the well can supply water for an extended period of time without lowering the level below the pump intake. That rate is considered the safe pumping rate for the well.

The information gathered during the test assists to determine the:

- Rate at which to pump the well
- Depth at which to place the pump.

The pump test must be performed at or above the pumping rate for which it is intended to operate. The test pump should be capable of pumping 125 percent of the desired yield of the well. Pumping should be continued at a uniform rate until the “cone of depression” reflects any boundary condition that could affect the performance of the well. This could be as short as six hours and as long as several days, depending on aquifer characteristics. The discharge rate and drawdown established should be maintained for specified time period. No pumping should be conducted at or near the test site for at least 24 hours prior to the test.

There are several accepted methods for determining aquifer properties. Which method should be used is dependent on the characteristics of the aquifer being tested (confined, unconfined, leaky confining layer) and the solution approach to be used. For an aquifer test, the well is pumped at a constant rate or with stepwise increased rates.

4.5.3. General Assumptions in Pumping Test

The underlying assumptions involved in the tests are:

- The aquifer is homogeneous, isotropic, and of uniform thickness over the area influenced by the test.
- The flow is Darcian
- Prior to pumping, the piezometric surface is horizontal, or nearly so, over the area to be influenced by the test.
- The well is fully penetrating, thereby the flow is horizontal to the well
- Well diameter is very small compared to radius of influence in aquifer
- The water removed from storage in the aquifer is discharged instantaneously with decline of head.
- Non-linear well losses are negligible.

The following assumptions/conditions apply to leaky confined aquifers:

- The aquitard is infinite in areal extent
- The aquitard is homogeneous, isotropic and of uniform thickness
- The water supplied by leakage from the aquitard is discharged instantaneously with decline in head

4.5.4. Constant Rate Test

In this method, the well is pumped at a constant rate typically for 12 hours to 7 days, while the water levels in the well are checked and recorded frequently as they decline from their standing water level to their pumping water level. Static water levels must be measured in advance of the test and after the test during recovery. Typically, pump tests shall be conducted for a minimum of 72 hours at a constant pumping rate. Time-drawdown *or* distance-drawdown approach can be used.

The volume of water pumped and the water level in the well are monitored simultaneously. Once the pumping has stopped, the recovery of the well (i.e., rising water level) is also monitored. Recovery measurements must be taken at specific time intervals (5 minutes for first 30 min. and 30 min. interval for the remainders) for several hours *or* until the water level returns to 90 percent of its original level. Using this data and applying certain calculations (described in a later section) we can determine the exact capacity of the well.

Observation Wells

For distance-drawdown approach, at least three observation wells (in the same line) should be monitored during the pump test. The horizontal distance between each observation well and the pumping well should be measured. The vertical elevation of a fixed reference point on each observation well and on the pumping well (e.g., “top of casing”) should be established. One observation well should be located outside of the expected influence of the pumping well; this observation well should serve to monitor background conditions during the pump test. The remaining observation wells should be placed so as to best define the hydrogeologic characteristics of the aquifer with respect to the pumping well.

For unconfined aquifers, at least two observation wells should generally be placed no farther than 100m from the pumping well and at least one additional observation well should be placed beyond the 100m radius. For thick confined aquifers that are considerably stratified, at least two observation wells should be placed within 250m of the pumping well and at least one observation well be located further than 250m from the pumping well.

Observation wells should be screened in, or open to, the same formation as the pumping well. Additional observation wells beyond the specified minimum number may be screened in, or open to, formations above or below the formation tapped by the pumping well to determine if there is any hydraulic connection between formations.

4.5.5. Step Wise Test

After the well development, the step-test should be performed by pumping at a minimum of 3, and preferably 4 rates, with the pumping time being three hours minimum at each rate. The highest rate should usually be at the highest rate at which the well was developed, with drawdown and flow rates recorded at minimum 5-minuteintervals during the first 30 minutes of each pumping rate, and at least 30-minute intervals for the remainder of each rate. Recovery readings should be taken at minimum 5-minuteintervals for the first 30 minutes after pump shutdown, and at least 30-minute intervals for four (4) to eight (8) hours.

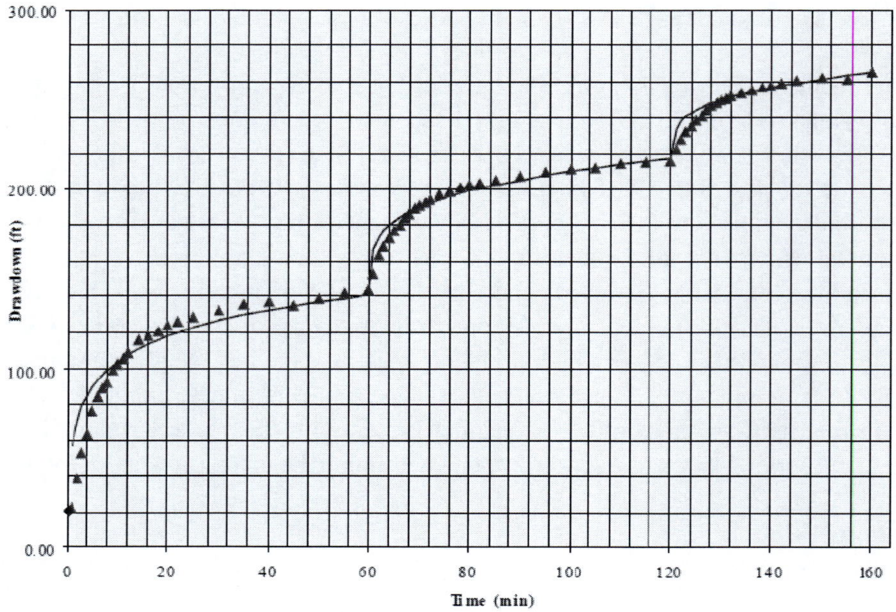

Figure 4.7. Schematic of Time drawdown plot for step-test.

After the step test is completed, a 24-hour pump test should be run at the rate calculated for the well based on the results of the step test. A schematic of time vs drawdown plot for step test is shown in **Figure 4.7.**

4.5.6. Analysis of Pump Test Data

The pump test and recovery data can be analyzed to determine the hydrogeological characteristics of the aquifer (such as hydraulic conductivity, transmissivity, and specific capacity) as well as to ascertain the eventual long-term production capability of the well.

Commonly used methods are:

- Theis drawdown
- Cooper-Jacob drawdown
- Theis recovery

Theis Method

Theis (1935) derived the following discharge-drawdown relationship, called Theis equation:

$$s = \frac{Q}{4\pi T} W(u) \tag{4.4}$$

Where,

$$W(u) = -0.577216 - \ln(u) + \sum_{i=1}^{\infty}(-1)^{i+1}\frac{u^i}{i.i!}$$

and,

$$u = \frac{r^2 S}{4Tt} \tag{4.5}$$

Where, *T* is the transmissivity of the aquifer, *t* is the time since pumping started, and *Q* is the pumping rate.

The steps in Theis curve analysis are as follows:

- Plot log drawdown (s) versus log time (t) on the same scale of matching (master) curve
- Overlay plotted and master curve (keeping axes parallel)
- Identify a match point and record values of s, t, u, and W(u)
- Solve for T as:

$$T = \frac{Q}{4\pi T}W(u)$$

- Solve for S as:

$$S = \frac{4uTt}{r^2}$$

- Derive K as:

$$K = \frac{T}{b}$$

Where, b is the thickness of the aquifer

Schematic of Thies curve is shown in Figure 8.

Cooper-Jacob Method

Cooper and Jacob (1946) observed that as the variables decreases with time ($u^i/i.i! \rightarrow 0$), and thus the term W(u) can be approximated by:

$$W(u) = -0.577216 - \ln(u)$$

The Theis equation then becomes:

$$s = \frac{Q}{4\pi T}\left(-0.5772 - ln\frac{r^2 S}{4Tt}\right) = \frac{Q}{4\pi T}\left(ln\frac{4Tt}{r^2 S} - 0.5772\right)$$

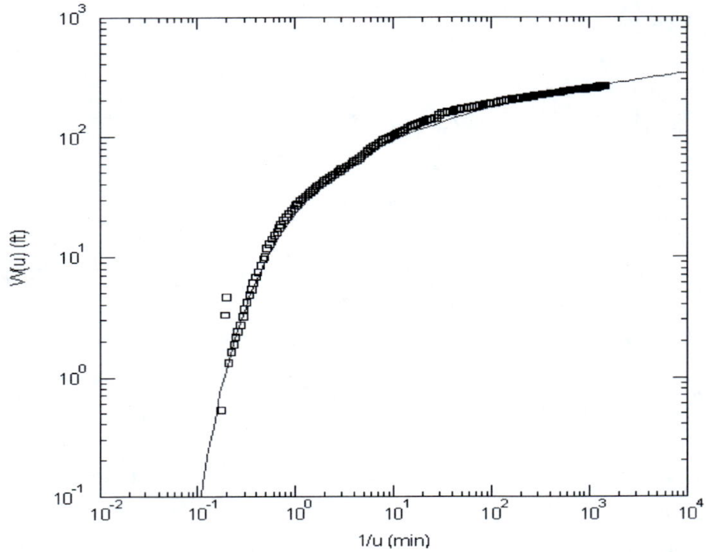

Figure 4.8(a). Well function vs 1/u (inverse of well parameter) curve (Thies master curve).

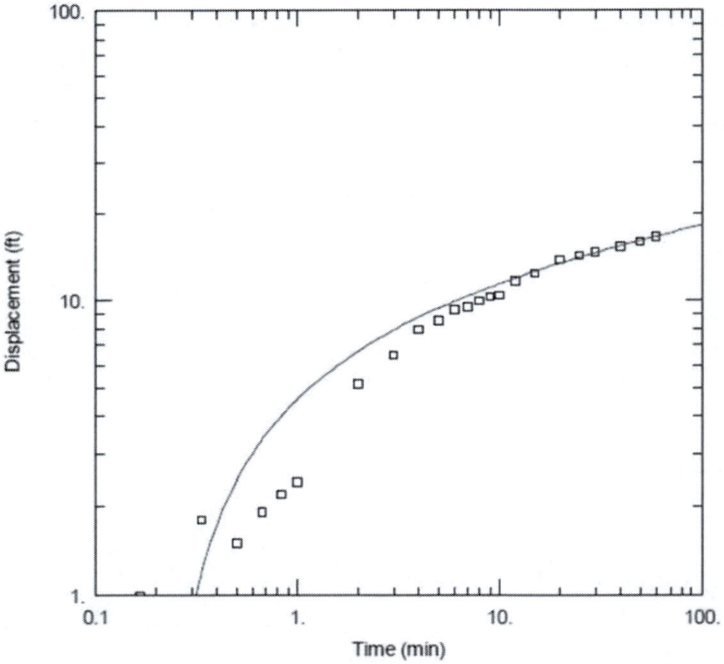

Figure 4.8(b). Theis analysis of time vs drawdown data.

In terms of $\log_{10}$ (As the term $\ln X = 2.303 \log_{10}X$), the above eqn. can be expressed as:

$$s = \frac{2.3Q}{4\pi T} log_{10}\left(\frac{2.25Tt}{r^2S}\right) \quad (4.6)$$

There are 3 types of solution approaches using Cooper-Jacob simplification:

- Time-drawdown
- Distance-drawdown, and
- Recovery

Time-Drawdown Approach

If the drawdown at time t_1 and t_2 are s_1 and s_2, the above eqn. can be written as:

$$s_2 - s_1 = \frac{2.3Q}{4\pi T} log_{10}\left(\frac{t_2}{t_1}\right)$$

If the data is plotted on semi-log paper, and the time t_1 and t_2 are chosen as one log-cycle apart, the above eqn. is reduces to:

$$\Delta s = \frac{2.3Q}{4\pi T}$$

Where, Δs is the drawdown per log-cycle of time. Or,

$$T = \frac{2.3Q}{4\pi\Delta s} \quad (4.7)$$

Again, from observation of eqn. (4.6), it is revealed that, when the term $log_{10}\left(\frac{2.25Tt}{r^2S}\right)$ is zero, the s value will be zero. That is,

$$log_{10}\left(\frac{2.25Tt}{r^2S}\right) = 0, s = 0$$

That is, when $\frac{2.25Tt}{r^2S} = 1$,

$$S = \frac{2.25Tt_0}{r^2} \quad (4.8)$$

The time for zero drawdown (t_0) can be obtained by extrapolating the graph to intersect the drawdown axis (**Figure 4.9**).

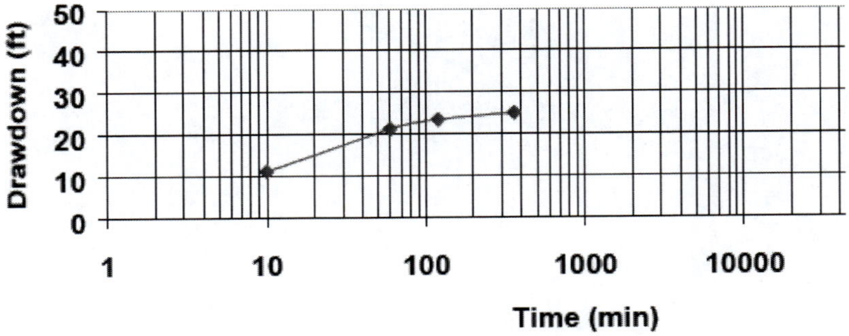

Figure 4.9. Schematic of Time-drawdown plot.

Distance-Drawdown Approach

If several observation wells are installed, then from the semi-log plot of distance-drawdown data, eqn. (4.6) can be written as:

$$s_2 - s_1 = \frac{2.3Q}{4\pi T} log_{10}\left(\frac{r_1}{r_2}\right)^2$$

Or,

$$\Delta s = \frac{2.3Q}{2\pi T} log_{10}\left(\frac{r_1}{r_2}\right) \quad (4.9)$$

Taking drawdown per log-cycle of distance in the graph (**Figure 4.10**), the T can be obtained as:

$T = \frac{2.3Q}{2\pi\Delta s}$, where Δs is the drawdown per log-cycle of distance (r).

Following the same principle of time-drawdown, the S value can be obtained as:

$$S = \frac{2.25Tt}{r_0^2} \quad (4.10)$$

Where, r_0 is the intersection of the straight-line portion of the graph with the abscissa at s = 0.

Theis Recovery Approach

Recovery test (also termed as residual drawdown test) involves measuring water level rise after the pump is shut down. In this approach, a well is pumped at a constant rate (Q) for a certain period of time (t). The residual drawdowns (s′) in pumped well *or* in an observation well are measured at different time intervals (t′) from the stop of pumping. The s′ and t/t′ are plotted in semi-log paper (s′ in log-scale) (**Figure 4.11**). The transmissivity T is obtained as that of Jacob's method as:

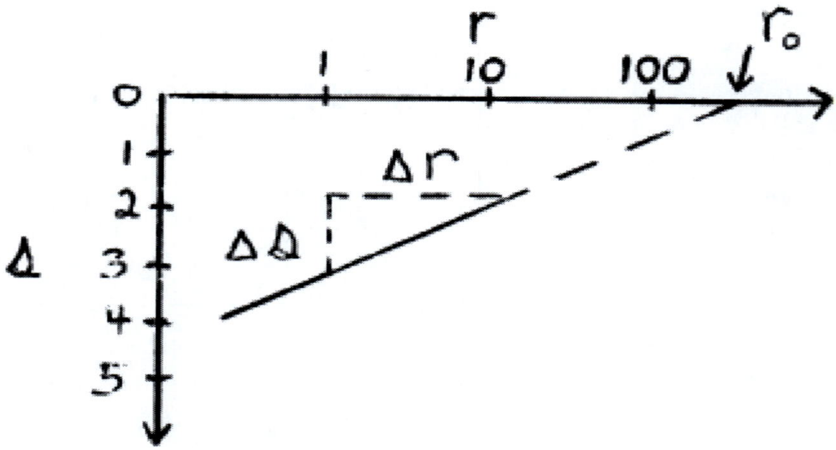

Figure 4.10. Schematic of distance-drawdown analysis.

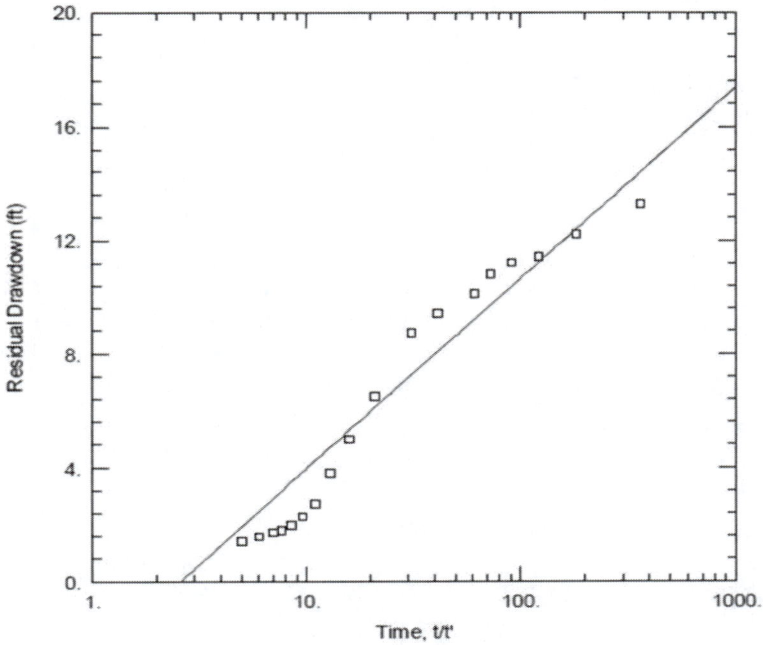

Figure 4.11. Graphical display of recovery data.

$$T = \frac{2.3Q}{4\pi\Delta s'} \tag{4.11}$$

The S can be determined from the data when pumping stopped, or from the time-drawdown data as:

$$S = \frac{2.25Tt_0}{r^2} \tag{4.12}$$

Limitations

In presence of hydraulic boundaries, the above method is not accurate.

Relevant Journals

Groundwater
Journal of Hydraulic Division – ASCE
Groundwater Hydrology
International J. of Applied Engineering and Technology
Hydrological Sciences Journal
International Journal of Engineering Science and Innovative Technology
International Journal of Research and Innovations in Earth Science
International Journal of Water Resources Development
Journal of American Water Resources Association
Research Journal of Environmental and Earth Sciences

Nature Geoscience
Journal of Geology & Geophysics
Journal of Hydrology
Water Science and Technology

Questions/Exercises

1. What are the types of well? Narrate the principal activities in well construction
2. Discuss the merits and demerits of different drilling methods
3. Briefly narrate the Design criteria of different elements of design
4. Discuss the importance of proper well completion
5. What are the well development methods? Briefly discuss.
6. Derive the equations of water flow into well for: (a) Confined aquifer, (b) Unconfined aquifer
7. Narrate the different methods of pumping test

References

Cooper, H. H. Jr. & Jacob, C. E. (1946). A generalized graphical method for evaluating formation constants and summarizing well-field history, Trans. *American Geophysical Union*,*27*(4), 526-534.

Dupuit, J. (1863). Etudes th_oriquesetpratiquessur le movement des eauxdans les canauxd_couverts et _ travers les terrains perm_ables [Theoretical and practical studies on the movement of water in open channels and through permeable rocks]. Editions Dunod Paris.

Theis, C. V. (1935). The relationship between the lowering of the piezometric surface and the rate and duration of discharge of a well using groundwater storage. *Transactions of the American Geophysical Union*,v. *16*, pp. 519-524.

Thiem, G. (1906). *HydrologischeMethoden*. Gebhardt (Leipzig).

Chapter 5

MANAGEMENT OF WATER RESOURCES

ABSTRACT

Along with energy, water is one of the major driving forces of economic development. Water has been identified as a crucial resource for all life, production, and development; while a lack of access to water has been linked to poverty. Although the natural water is adequate in absolute terms, it is not available at the right time, place, quantity and quality. Competition in limited water resources increasingly occurs among agriculture, rural, urban, industrial and environmental uses. It is desired to use water in more efficient, equitable, and environmentally sustainable ways. Improved and efficient water management practices can help to maintain profitability in an era of increasingly limited and more costly water supplies.

5.1. CONCEPT OF WATER RESOURCES MANAGEMENT

According to Griffin (1997), "Management is a set of activities (including planning and decision making, organizing, leading, and control) directed at an organization's resources (e.g., human, financial, physical, information) with the aim of achieving organizational goals in an efficient and effective manner."

Water management is concerned with how the available supply of water can be better allocated and utilized to fulfill the command. It includes the intervention of humans in the manner in which surface and/or ground water is captured, conveyed, utilized and drained in a certain area.

Water should be regarded as an economic good does not necessarily imply that a 'market price' must be paid for it, or even must be paid for at all. It means simply that water is a scarce and valuable resource that should not be wasted.

5.1.1. Goals of Water Resources Management

The main purpose of water resources management is to meet needs of human and nature. Broader goals and objectives of water resources management are:

- welfare of human being and improve the quality of life
- national/regional economic development
- income distribution
- ecosystem services
- health and safety
- cultural and educational opportunities

Water resources should be managed in the context of a national water strategy that reflects the nation's social, economic, and environmental objectives and is based on an assessment of the country's water resources.

In water resources management, two fundamental aspects are to be considered:

(i) Planning
(ii) The implementation process

5.2. Strategies for Water Resources Management

Strategies for water resources management can be categorized as:

(a) Demand-side management
(b) Supply-side management
(c) Conjunctive use of surface and groundwater

5.2.1. Demand-Side Management

5.2.1.1. Concepts

Demand management options refer the actions that influence the use of water after the entry point. Demand side management is commonly implemented together with a water conservation program. Water conservation is generally accepted to mean "the minimization of loss or waste, the preservation, care and protection of water resources and the efficient and effective use of water."

Demand side management involves a broad range of measures that aim to increase the efficiency of water use. Demand management may be defined as the adaptation and implementation of a strategy by a water institution to influence the water demand and usage in order to meet any of the following objectives: economic efficiency, social development, social equity, environmental protection, sustainability of water supply and services and political acceptability.

5.2.1.2. Different Approaches of Demand Management

The measures for demand management can include:

- conservation-oriented rate structures
- leak reduction program

- landscaping with drought tolerant species
- water savings irrigation practices in agriculture
- water reuse
- pressure management
- water audit
- public awareness campaigns, etc.

Demand side management can facilitate to avoid construction of new water or wastewater infrastructure; decrease operation and maintenance costs; and decrease environmental impacts of increasing withdrawals.

Techniques for urban sector may include:

- water loss management
- escalating block rate tariffs
- promotion of water-wise industries
- water auditing
- retrofitting with water-saving devices
- water-wise gardening
- public awareness

Techniques in the agricultural water sector can include:

- efficient/water-saving irrigation scheduling
    - alternate furrow irrigation
    - alternate wetting & drying of rice field
    - soil drying during ripening of crop
    - deficit irrigation
    - adopting efficient irrigation method
- reduction/removing of pricing subsidies
- Soil-water conservation measures
    - Reducing evaporation
    - mulching by various elements
    - using super absorbents
- low water-demanding crops
- appropriate cropping pattern
- adopting efficient irrigation devices/technologies
- land leveling
- water-wise tilling/field preparation
- water-wise crop cultivation method
- soil-crop-weather management
- Ponding rain-water at rice field by high levees (bunds)
- Storing rainwater in farm pond, lake, etc. and use in dry period

In demand side measures, there may be considerable variations in the water savings among the techniques. The appropriateness of different types of demand side measures may be a function of factors related to locality, value of water, financial capacity and habitat of the farmers, etc. Any combination of the measures or awareness campaign may be useful in a particular area, and hence can be adopted.

5.2.2. Supply-Side Management

Supply management options refer the actions that affect the quantity and quality of water at the entry point of the distribution system.

5.2.2.1. Methods or Approaches of Supply Management

Supply management options can include:

(i) Development of new supply source
   i. surface water (barrage, dam, artificial canal, rubber dam, etc.)
   ii. groundwater

(ii) Import of virtual water
(iii) Rainwater harvesting at farm pond, lakes
(iv) Augmenting natural recharge (for increasing groundwater)
(v) Inducing artificial recharge (for increasing groundwater)
(vi) Sharing of international rivers
(vii) Use of non-conventional water with appropriate strategic measures
   - saline water/poor quality water
     - with mixing/blending ratio
     - applied poor quality water at non-sensitive growth stages
   - irrigation effluent
   - drainage outflow

(viii) Network rehabilitation
(ix) Reallocation through volumetric constraints
(x) Use of super absorbents
(xi) Implementing dual water supply system
(xii) Implementing duel water use – aquaculture in rice field

5.3. Sustainable Yield

There are several ways to define sustainable yield. As an example, the definition of sustainable yield proposed by the National Groundwater Committee of Agriculture and Resource Management Council of Australia and New Zealand, is as follows:

Sustainable yield is the groundwater extraction regime, measured over a specified planning timeframe that allows acceptable levels of stress and protects the higher value uses that have a dependency on water.

An estimated volume of sustainable yield of groundwater is the key step in a process that, for high priority aquifers, can be complex. In one sense, the rigour of undertaking the process to arrive at the sustainable yield is as valuable as the final number itself. In essence, sustainable yield is not just a number; it is as much the concept of deriving a working knowledge of the aquifer in question.

Once the sustainable (or acceptable) yield is defined then a conjunctive resource management plan can be developed and implemented. The key elements of a plan are the operating rules, policy framework, targets and monitoring and evaluation strategy.

There are two guiding principles: access and trading, and the precautionary approach to implementation of conjunctive management principles. Whilst the process can be rigorous, the outcomes, however, can never be certain. There will always be technical errors in the system either inflicted by uncertainties within the analytical processes used in estimating the water balance, or by unforseen aspects of the natural environment. Then there are the social and economic consequences that need to be fully considered when developing plans. These aspects may mean that the best technical initiatives are delayed or not adopted. It is of paramount importance, that, having quantified user provisions and allocated resources according to a predictive framework, to monitor and evaluate the system to ensure there is in fact convergence on the targets specified within the specified timeframes.

5.4. Climate Change and Water Resources

Climate change is likely to have significant impacts on the availability of fresh water. Climate change will affect water resources through its impact on the quantity, variability, timing, form, and intensity of precipitation. This paper provides an overview of the projected physical and economic effects of climate change on water resources in North America (with a focus on water shortages), and a brief discussion of potential means to mitigate adverse consequences. Additional effects of global climate change that have important implications for water resources include increased evaporation rates, a higher proportion of precipitation received as rain, rather than snow, earlier and shorter runoff seasons, increased water temperatures, and decreased water quality in both inland and coastal areas. The physical and economic consequences of each of these effects are discussed below.

Increased evaporation rates are expected to reduce water supplies in many regions. The greatest deficits are expected to occur in the summer, leading to decreased soil moisture levels and more frequent and severe agricultural drought. More frequent and severe droughts arising from climate change will have serious management implications for water resource users. Agricultural producers and urban areas are particularly vulnerable, as evidenced by recent prolonged droughts in the western and southern United States, which are estimated to have caused over $6 billion in damages to the agricultural and municipal sectors. Such droughts also impose costs in terms of wildfires, both in terms of control costs and lost timber and related resources.

Rising surface temperatures are expected to increase the proportion of winter precipitation received as rain, with a declining proportion arriving in the form of snow. Snow pack levels are also expected to form later in the winter, accumulate in smaller quantities, and melt earlier in the sea son, leading to reduced summer flows. Such shifts in the form and timing of precipitation and runoff, specifically in snow-fed basins, are likely to cause more frequent summer droughts. Research shows that these changes are already taking place in the western United States.

Changes in snow pack and runoff are of concern to water managers in a number of settings, including hydropower generation, irrigated agriculture, urban water supply, flood protection and commercial and recreational fishing. Timing of runoff will affect the value of hydropower potential in some basins if peak water run-off occurs during nonpeak electricity demand. Energy shortages and resulting energy price increases will provide incentives to expand reservoir capacities or develop alternative energy sources. A shift in stream hydrographs to more winter flow may also disrupt the life cycle of cold water fish species, such as salmon, which depend on late spring flows to "flush" young salmon to the ocean, and on summer flows to moderate water temperatures. Unless winter runoff is captured and stored for late spring or summer use, fewer salmon smolt will survive migration and more frequent fish kills will occur from lethal stream water temperatures. Climate change is expected to affect water quality in both inland and coastal areas. Specifically, precipitation is expected to occur more frequently via high-intensity rainfall events, causing increased runoff and erosion. More sediments and chemical runoff will therefore be transported into streams and groundwater systems, impairing water quality.

Water quality may be further impaired if decreases in water supply cause nutrients and contaminants to become more concentrated. Rising air and water temperatures will also impact water quality by increasing primary production, organic matter decomposition, and nutrient cycling rates in lakes and streams, resulting in lower dissolved oxygen levels. Lakes and wetlands associated with return flows from irrigated agriculture are of particular concern. This suite of water quality effects will increase the number of water bodies in violation of today's water quality standards, worsen the quality of water bodies that are currently in violation, and ultimately increase the cost of meeting current water quality goals for both consumptive and environmental purposes. Rising sea levels could also reduce water quality and availability in coastal areas. Recent projections of sea-level rise by the end of the 21st century range from 19 to 58 cm. A more dramatic increase in sea-level, on the order of meters rather than centimeters, is possible, but most scientists consider it a low probability risk. For example, complete melting of the Greenland Ice Sheet or West Antarctic Ice Sheet would trigger such a large rise. Rising sea levels could affect groundwater quality directly via saltwater intrusion. Radical changes to the freshwater hydrology of coastal areas, caused by saltwater intrusion, would threaten many coastal regions' freshwater supplies.

Rising sea levels could also affect water availability in coastal areas indirectly by causing water tables in groundwater aquifers to rise, which could increase surface runoff at the expense of aquifer recharge. Water shortages will cause the price of water to rise, through monthly water bills or one-time connection fees for new homes and businesses. A sufficiently large price increase could affect the extent and pattern of urban growth throughout the United States. Costly water supply projects, such as desalination plants, pipelines, and dams will also become more economically attractive.

One final and important effect of the water resource impacts discussed above is the potential for more frequent and intense interstate and international water allocation conflicts. Water markets have the potential to prevent or diffuse such conflicts; however, the assignment of water rights to establish the market can create more conflict than it diffuses.

Effective operation of water resources systems requires quantification of regional climatic variability. The frequency and magnitude of critical wet and dry periods must be understood in order to provide a stable and reliable source of water. Recent advances in the understanding of the influence of hydroclimatic events on hydrology have led to the use of climate information in water resources planning. An analysis of an innovative approach to utilizing climate information in the quantification of regional hydrological variability for water resources planning is presented using a dynamic hydrological system response probability distribution that is conditioned upon a specified event in the hydroclimatic system.

Impacts of Climate Change on the Water Cycle

Climate and water variability significantly impact all aspects of our daily lives and socioeconomic development. Any modification can bring about important consequences with respect to water availability, health, food, tourism, forestry, transportation, and habitat of living species, also affecting societies and their ability to develop in a sustainable way Approximately 90% of all natural disasters worldwide are due to meteorological, weather or hydrological conditions. Life and material losses caused by natural disasters are a major obstacle to sustainable development and security. Therefore, it is essential to compile accurate data, observations and forecasts and ensure the timely and reliable exchange of information on weather, climate and water.

The impact of climate change on water resources depends on the changes in volume and quality of surface and groundwater, wastewater discharges, riverbeds, and the size of human settlements as well as water demands for human uses. Therefore, measuring the hydrological cycle provides essential information to be used in early-warning systems to reduce negative impacts or to restore watersheds, aquifers and ecosystems. Measurements of hydrological variables through radar, meteorological observatories, weather and gauge stations and groundwater wells are helpful in assessing the quantity and quality of surface and groundwater, which is essential to protect life, property and the environment.

Impacts on the hydrological cycle are reflected in an increase in temperature and changes in precipitation, as these affect evaporation from seas and oceans and transpiration from plants, runoff of watersheds, and the infiltration that recharges aquifers. At present, evidence and data regarding the magnitude and extent of these variations have proven hard to gather. However, it is certain that climate change will affect water availability and increase water demand for most uses, for instance in ecosystems and agriculture. Demand may exceed supply, creating major disasters in vulnerable areas. Also, there will be differentiated changes in water quality in rivers, lakes, wetlands and coastal ecosystems. It is thus important that countries monitor these changes, a step towards sustainable WRM.

5.5. Conjunctive Use of Surface and Groundwater

5.5.1. The Concept of Conjunctive Use

A critical problem that mankind has to face and cope with is how to manage the intensifying competition for water among the expanding urban centres, the agricultural sector and in-stream water uses dictated by environmental concerns. Confronted with the prospect of heightened competition for available water and the increased difficulties in constructing new large-scale water plants, water planners must depend more and more on better management of existing projects through basin-wide strategies that include integrated utilization of surface and groundwater (conjunctive use).

In general terms, conjunctive use implies the planned and coordinated management of surface and groundwater, so as to maximize the efficient use of total water resources. Because of the interrelationship existing between surface and subsurface water, it is possible to store during critical periods the surplus of one to tide over the deficit of the other. Thus groundwater may be used to supplement surface water supplies, to cope with peak demands for municipal and irrigation purposes, or to meet deficits in years of low rainfall. On the other hand, surplus surface water may be used in overdraft areas to increase the groundwater storage by artificial recharge. Moreover, surface water, groundwater or both, depending on the surplus available, can be moved from water-plentiful to water-deficit areas through canals and other distribution systems. On the whole the integrated system, correctly managed, will yield more water at more economic rates than separately managed surface and groundwater systems.

The current trend in aquifer management focuses on determining the maximum and minimum water levels, in order to regulate storage capacity. As a matter of fact, uncontrolled overexploitation causing progressive drawdown below the minimum permissible piezometric levels, will lead to increased pumping costs, land subsidence, infiltration of poor quality water, drying up of springs and shallow wells, and decreased river flows. Moreover, in coastal aquifers the prolonged reduction of freshwater flow towards the sea reduces the equilibrium gradient, inducing saltwater intrusion and the inland movement of the freshwater–saltwater interface. Combining so many aspects requires methods of analysis that systematically integrate them in such a way that within the planning process alternative solutions can be defined, tested and chosen.

Normally artificial groundwater recharge is accomplished by means of infiltration basins or injection wells. Other techniques for augmenting subsurface supplies include vegetation management, runoff inducement and increasing seepage from streams by widening the wetted perimeter of channel sections or lowering the groundwater table in the floodplain. Water quality aspects play a major role in this process. They mainly concern the quality of recharge water and its effects on groundwater quality. Generally speaking, numerous issues need to be addressed before suitable recharge systems can be chosen, designed and managed for optimum environmental and economic performance.

5.5.2. Framework for Conjunctive Resource Management – Guiding Principles

Identification of Resource Management Issues

The first part of the process involves the identification of conjunctive resource management issues that need to be addressed. The issues may be related to protection of the resource for consumers or it may be an issue associated with protection of water for dependent ecosystems. Identification of the key issues at the start of the process creates focus for the assessments and helps set objectives and obtains agreement on the desired outcomes amongst the community and other stakeholders. Examples of issues that may be significant include:

- Protection of baseflow;
- Access to flood flows;
- Salinity management;
- Achieving equity between water users.

While most of the key issues will be identified at the start of the process, these are likely to change as the assessment and evaluation process progresses. Stakeholder consultation during the identification of the key issues is recommended.

Identification and Quantification of Water Users and Uses

The next step in the process involves identification and quantification of water users and uses. The level of certainty in an estimate of water use will vary considerably. For example, irrigation use is often metered and relatively easy to quantify, but environmental water requirements (EWR) are usually unknown or not expressed volumetrically. Often, water use information is only available after a water plan has been implemented, and so use of this information is normally part of the performance monitoring or review process.

The availability of information on water use (both consumptive and environmental) will directly affect the reliability of the technical assessments and definition of user provisions.

Confirmation of the (External) Decision Environment

An understanding of the external decision making environment is needed to understand the opportunities and constraints provided by existing legislation and planning policy. For example, it will be difficult to implement detailed operating rules where the taking of a water resource is unregulated. The context of water management in an area where there are well-developed strategies for integrated catchment management (ICM) needs to be described and considered to avoid potential conflict. For example, an ICM strategy for revegetation for dryland salinity control could be in conflict with maintenance of a groundwater resource for consumptive use if the re-vegetation strategy targets areas with high recharge potential. There may also be instances where particular features of a catchment are non-negotiable and will not be traded at any cost, such as a world heritage listed wetland.

In addition, there may be substantial social and economic factors that need to be considered in determining sustainable yield and conjunctive use approaches in plans. To fully assess these aspects specialist studies may be required to complement the technical and environmental studies.

Technical Assessments

The technical assessments form a major part of the process with the intention being to estimate the sustainable yield and provide information for the definition of user provisions. The outcome from definition of user provisions is agreement to the sustainable yield that provides equity amongst users. The four main components of the technical assessments process are knowledge generation, conceptualisation, modelling the system behaviour and reporting. The importance of peer review and quality assurance at this point in the process is emphasised. This part of the process also involves consideration of the costs/impacts of using groundwater under the "no-plan scenario" (such as ecosystem impacts) and benefits/opportunities from developing a groundwater resource, such as the economic development that it can bring. The costs and benefits analysis is not often undertaken explicitly in water planning, but can help to assess the implications of trade-offs undertaken when defining user provisions.

There are several guiding principles that need to be considered at this point in the process. The main guiding principle governing the technical approach is the need to assess surface water and groundwater as "one resource." Other guiding principles relate to aquifer, and stream-aquifer interaction classification, performance and monitoring indicators, minimum datasets understanding uncertainty and variability, reporting and management of risk.

Surface Water and Groundwater (One-Resource) Balances

The cornerstone to the assessments process is the recognition that surface water and groundwater are two inter-related components of the one resource, and changes in one system can have a significant influence on the other. However, one resource should not dominate over the other during the assessment process. Therefore, ideally the sustainability issues of both the surface and groundwater resources of a catchment need to be assessed together. Although this is rarely undertaken at present, it is achievable because the same basic analytical approaches and tools (models) are involved in the assessment of each resource. The goal is to conjunctively assess and manage the one resource for sustainability over a specified management timeframe, avoiding allocation of the resource twice (from "separate" pools of surface and groundwater) and minimising impacts from irrigation. The conjunctive assessment and management of water resources involves acknowledging the different attributes of surface and groundwater systems within management approaches to obtain maximum benefit (within a specified timeframe) for economic, environmental and social values.

The integration of knowledge of surface water and groundwater at the assessments stage allows the development and testing of conjunctive use strategies such as aquifer storage and recovery, protection of ecosystems and opportunistic cycling of the use of each resource within wet and dry periods. It is considered that the framework should be applied not only to water resources management in systems where there are connections between aquifers and streams, but also in systems that are disconnected. It has been shown that even in disconnected surface-groundwater systems; the use of one resource can affect the other. Disconnected stream-groundwater areas tend to be associated with unregulated stream sections or mid to lower alluvial areas of catchments. The connected recharge and discharge areas may be distant but should not be ignored in the water management planning.

The framework emphasises that all groundwater and surface water systems have some degree of effective connection, thus the conjunctive management of "the one resource" should be adopted, especially where the aquifer system is not robust.

Investigations

There are four main areas of investigation that need to occur as part of the assessment process, they are:

- Knowledge generation;
- Conceptualisation;
- Modelling system behaviour; and
- Reporting.

Knowledge generation aims to provide an understanding of how the groundwater system works which provides the information required to develop the conceptual models of the hydrogeological and hydrological system. Typically knowledge generation involves collection of data to define the spatial distribution of aquifers, to characterise aquifer hydraulic properties, estimate recharge and discharge, to quantify surface water – groundwater interactions and to develop the pre- and postwater balances.

The conceptual model establishes the architecture or framework for the predictive analysis by explaining the nature and behaviour of groundwater systems, and the way in which systems are likely to respond both spatially and over time. All assessments must be underpinned by an analysis of the entire aquifer system of interest, not just the lowest salinity resource areas or an administrative region imposed as a result of where the groundwater users are located. A conceptual model that includes all elements of the biophysical system within a dynamic water balance framework must support this analysis. In higher-priority aquifers, the analysis should be based on distributed parameter predictive mathematical models.

Outcomes from the conceptualisation process include:

- classification of the linkages between surface and groundwater systems, and identification of the potential for impacts;
- conceptualisation of the systems in practical terms, detailing the flow and water quality processes involved so that detailed models can be developed;
- a check on the applicability of the conceptualisation by doing simple calculations such as water balances, which require information on environmental water requirements (spatially and temporally); and
- prioritisation of further investigations.

Within the conceptualisation stage, there should be sufficient understanding of the system to develop preliminary bounds on the major factors that restrict the combined sustainable yield of the surface and groundwater systems. Such bounds may relate to:

- the maximum extent to which groundwater levels could be permitted to change, based on resource, environmental or economic reasons;
- the minimum environmental water provision constraints;
- the minimum volumes that need to be diverted to maintain social and economic viability in the community; and/or
- others of specific interest to the catchment or community.

A predictive tool is required to guide the development of management rules (see Defining User Provisions) that allow the water resources to be put under the maximum acceptable levels of stress while still protecting the higher value uses that have a dependency on the water. Computer-based mathematical modelling is the only tool that has been identified as having this capability. Resource managers can use mathematical models to quantify the effects of trialling various management options, simulate various use and climatic variability scenarios, and demonstrate the findings to the community. The modelling methods that should be used for conjunctive resource management depend on the data available for setting up the model and for its calibration, the complexity of the stream aquifer interaction process, the significance and environmental sensitivity of the water system, the economic importance of the area and the study resources available.

For complex linked systems with high sensitivities and values, more complex models need to be used. MODFLOW is recognised as the world standard groundwater flow model, and has several surface water linkage modules that could be utilised. However, as surface water resources are of major importance, catchment models for most stream systems have been, or are being, developed. The process of obtaining estimates of sustainable yield should be well documented, with emphasis placed on describing the conceptual model, nature of the data used, the techniques employed and the assumptions made. This reporting also should be made available publicly, and should undergo peer-review for the more significant aquifer systems.

Assessing Impacts from the Use of Groundwater

The framework highlights the importance of understanding the costs and benefits associated with the extraction and use of groundwater. Within an ICM framework all water using activities may have a right to access water and there will be costs and benefits associated with each use. The definition of user provisions involves consideration of the trade-offs between water users and the costs and benefits each user brings.

One of the costs (or potential impacts) associated with the use of groundwater for irrigation is the change in the water and salt balance that can lead to problems such as groundwater mounding, waterlogging and salinisation of soils, and increased discharge of salt to the riverine environment. A key reason for including this analysis in the conjunctive management framework is that the use of water for irrigation may reduce the capacity of the groundwater resource (or surface water resource) to meet future demand, that is reduce the sustainable yield. For example, groundwater mounding within a shallow more saline aquifer will induce downward leakage of salt to the less saline productive aquifer. Other impacts could relate to aquifer storage depletion, and if excessive, land subsidence and infrastructure impacts. While these impacts are unlikely in most of the groundwater systems in the basin, the potential should be assessed in over allocated and over used groundwater systems.

5.6. Integrated Water Resources Management

Integrated water resource management depends on cooperation and partnerships at all levels, from individual to governmental and non-governmental, national and international organizations sharing a common political, scientific and ethical commitment to the need for water security and for optimizing water resources use and planning. To achieve this goal, there is a need for coherent national, regional or interregional policies to overcome fragmentation, and for transparent and accountable institutions at all levels. To this end, targets should be established and suitable strategies should be devised to meet the challenges inherent in the sustainable use and development of water resources. These resources should be managed at both the river basin and aquifer levels. The management framework should envisage a high level of autonomy for the body responsible for water use and planning, allow for stakeholder participation in decision-making and generate and disseminate information. Where appropriate, specific river basin, catchment and groundwater authorities should be set up, and their capacities enhanced. Where water is shared, actions should be taken to build confidence among riparian states, enabling them to accept some form of restricted sovereignty regarding their common resource, based on both equitable utilization and regional cooperation. Besides institutional strengthening, sound and fair financial management, based on the "user pays" principle, is needed to improve the efficiency of services, provide additional resources for investment, encourage demand management, and promote pollution control and prevention.

5.7. Water Management and the Environment

Water resource management should preserve or enhance the environment's buffering capacity to withstand unexpected stress or negative long-term trends. As the environment's carrying capacity is put under increasing pressure, due to the growing needs of the population and improper use of its resources, environmental vulnerability increases too. In this context, mismanagement of water resources, paying only lip service to the environment, has led to water scarcity and water pollution which threaten the security and the quality of human life.

Ecosystem Services

The well-being of present and future generations depends on access to and the availability of ecosystem services. *Ecosystem services* are the benefits people obtain from ecosystems. These include provisioning services such as food, water, timber, and fibre; regulating services that affect climate, floods, disease, wastes, and water quality; cultural services that provide recreational, aesthetic, and spiritual benefits; and supporting services such as soil formation, photosynthesis, and nutrient cycling.

Ecosystems provide not only the basic materials that society needs to survive, including food, but they also foster less obvious measures such as health, good social relations, security, well-being and the freedom to decide on our own lives. In order to achieve social and economic development goals, ecosystem users and managers often make trade-offs between the available services, particularly as the use of resources has begun to outstrip their supply. Managers need to make decisions between different services, whether to reap benefits now or retain them for future use and how to meet the needs of both ecosystems and society.

Decisions regarding trade-offs are complicated by the perceived value of ecosystem services by different stakeholder groups. For example, an upstream industry may value a river system as a conduit to carry away waste while downstream users enjoy the same system for recreational purposes. Furthermore, supporting and regulatory services are often ignored by policy makers as they depend on slowly changing factors which are difficult to quantify and monitor. The lack of attention afforded to regulatory and supporting services may result in a lack of resilience and render socio-ecological systems more vulnerable to environmental change. Identifying trade-offs enables decision makers to understand the consequences of favouring one ecosystem service over another, which is essential if ecosystems services are to be managed to maximise human well-being.

Sustainable Yield

There are several ways to define sustainable yield. As an example, the definition of sustainable yield proposed by the National Groundwater Committee of Agriculture and Resource Management Council of Australia and New Zealand, is as follows:

Sustainable yield is the groundwater extraction regime, measured over a specified planning timeframe that allows acceptable levels of stress and protects the higher value uses that have a dependency on water.

An estimated volume of sustainable yield of groundwater is the key step in a process that, for high priority aquifers, can be complex. In one sense, the rigour of undertaking the process to arrive at the sustainable yield is as valuable as the final number itself. In essence, sustainable yield is not just a number; it is as much the concept of deriving a working knowledge of the aquifer in question.

Once the sustainable (or acceptable) yield is defined then a conjunctive resource management plan can be developed and implemented. The key elements of a plan are the operating rules, policy framework, targets and monitoring and evaluation strategy.

There are two guiding principles presented here: access and trading, and the precautionary approach to implementation of conjunctive management principles. Whilst the process can be rigorous, the outcomes, however, can never be certain. There will always be technical errors in the system either inflicted by uncertainties within the analytical processes used in estimating the water balance, or by unforseen aspects of the natural environment. Then there are the social and economic consequences that need to be fully considered when developing plans. These aspects may mean that the best technical initiatives are delayed or not adopted. It is of paramount importance, that, having quantified user provisions and allocated resources according to a predictive framework, to monitor and evaluate the system to ensure there is in fact convergence on the targets specified within the specified timeframes.

5.8. Planning and Management of Shared Waters

Almost all waters across the globe are shared, but some are shared more than the others. Water is jointly used at all scales of society, such as individual households, villages, districts, states, countries and continents. Depending upon the nature and extent of influencing factors, some shared waters havemore significance for individuals and communities than the others. Such influencing factors often include, but are not limited to: (1) the number of people sharing the water source, compared to availability (in quantity and quality); (2) the different competing demands for water (e.g., domestic, agriculture, industry, recreation and the ecosystem); and (3) the social, economic, political, cultural, racial, religious and other differences between the groups sharing the water resource. Politically, when shared waters cross national boundaries, the stakes often get much higher, as nationalism may play a key role in their management (and negotiations regarding their management) in addition to concerns regarding human safety and security, economics, and sectoral survival. This is why issues associated with transboundary waters (i.e., water crossing national boundaries) often attract far more attention in the academic, scientific and other literature, and the media, although their enormous sizes and the extent of demands they can meet are also obvious reasons.

Currently, about 275 river basins and about 270 groundwater aquifers are shared by two or more nations (UNESCO, 2009). All together, these transboundary waters cover about half of the earth's surface and serve about 40% of the world's population. Such bodies have been sources of numerous conflicts between the nations sharing them – and subjects of significant cooperation at various levels of governments and in various water sectors. In some instances, water is made to play a role in conflict or cooperation between nations even when it is not shared across geographical borders. Regardless of whether there is conflict or cooperation over transboundary waters, it is fair to say that such waters are, in general, not adequately planned, developed, and managed.

There are many international water frameworks and regulations, and numerous bilateral and multilateral agreements and treaties between nations, for sharing transboundary waters (Salman, 2007). However, a variety of factors influence their formulation and implementation, often in very complex and interconnected ways, and thus present additional challenges to the management of these waters. In almost all transboundary waters, the influencing factors and the associated challenges are related to both "hard" and "soft" sciences, and include climate and water science, engineering and technology, economics, society, politics, education, environment and law (Sivakumar, 2013). Although the extent to which one given factor or discipline influences how shared waters are managed often varies from one case to another, politics is perhaps the dominant factor, especially when it drives or is driven by nationalism.

Notwithstanding the progress made, the enormous difficulties that continue to be encountered in the planning and management of transboundary and other shared waters clearly indicate the influential roles of such non-water factors. The conflicts associated with the Jordan River in the Middle East, the Nile River in Africa, the Indus River and the Ganges-Brahmaputra-Meghna river system in South Asia, and the Parana´ River in South America are just a few examples.

Politics and human behaviour play important roles in the planning and management of water resources that are shared across all scales of society. In recent years, there have been significant advances in studying these roles, especially through the creation of the fields of hydropolitics and hydropsychology. However, the way in which these roles have been represented in studies thus far and the underlying assumptions therein present inherent difficulties and limitations in dealing with water planning and management across all scales. In our dynamically linked and constantly evolving human society and, consequently, in our water demands and activities for a comprehensive account of the changing global water management landscape, a reconciliation of hydropolitics and hydropsychology can also offer new avenues for dealing with some important water concepts and conflicts, such as real and virtual water trading, water pricing, river linking schemes, large dams, water agreements and legislation, and climate change adaptation and mitigation. Although the applications are no doubt universal, they are particularly relevant for developing countries, where the combined influences of politics and human behaviour on water planning and management, and vice versa, are becoming increasingly important. They should also greatly help developing democracies, where politics and human behaviour are deliberated in a more transparent manner than ever before. It is hoped that the preliminary attempt made in this study in favour of hydropolitics–hydropsychology reconciliation will lead to a much stronger qualitative and quantitative assessment of the framework in the future.

5.9. Water Law and Policy

Water - as a natural resource with both ecological and economic value - is of vital importance for sustaining life, health and the integrity of ecosystems. This resource is, however, increasingly threatened by over-use and insidious long-term effects of pollution. Pollution comes from both point sources and diffuse sources. Potential risks or actual impacts could permanently impair underground water resources, with far-reaching and unpredictable implications for present and future generations. The *Charter on Water Management* provides policy measures for such action. The governments should apply the provisions of the *Charter* when formulating, adopting and implementing water-related policies and strategies at both national and international levels.

Ground-Water Policy

Governments should formulate and adopt a long-term policy to protect ground water by preventing pollution and overuse. This policy should be comprehensive and implemented at all appropriate levels. It should be consistent with other water-management policies and be duly taken into account in other sectorial policies.

Ground-Water Strategies

1. As ground water should be recognized as a natural resource with economic and ecological value, ground-water strategies should aim at the sustainable use of ground water and preservation of its quality. These strategies should be flexible so as to respond to changing conditions and various regional and local situations.
2. Ground-water pollution is interrelated with the pollution of other environmental media (surface water, soils, atmosphere). Ground-water protection planning should be incorporated into general environmental protection planning.
3. Protection measures aimed at prevention of ground-water pollution and over-use should be the basic tools for ground-water management. Such protection measures include, *inter alia,* monitoring of ground waters, development of aquifer vulnerability maps, regulations for industry and waste disposal sites paying due account to ground-water protection considerations, geo-ecological assessment of the impact of industrial and agricultural activities on ground waters, and zoning of ground-water protection areas.

Groundwater Legislation

1. Provisions of the Charter should be applied in national ground-water legislation. Legal provisions specific to peculiarities of ground-water management should be formulated and promulgated. Legislation should contain provisions for its effective implementation including the mandate, competence and power of the relevant authorities in accordance with uniform principles, e.g., as set out in this document. Governments' rights to control ground-water abstraction and use as well as all activities with a potential impact on the quantity and quality of ground-water resources should be recognized by legislation.
2. Adequate definitions of ground-water characteristics, use and protection should be formulated and integrated into legislation with a view to avoiding ambiguities and thus facilitate implementation of legal provisions for ground-water management.
3. Ownership with respect to ground water should be clearly defined in a water act or code depending on national legislation. Ground water should be declared in the public domain or authority should be vested in Government to restrict, in the public interest, the rights accruing from its private ownership. New legislation should strive towards changing ownership rights into ground-water use rights subject to a government-controlled permit system. To this end, it would be necessary to draw up precise rules concerning the selection of criteria applicable for the recognition of ground-water use rights and for the granting of permits taking into account orders of priority for the allocation of available water. Such rules should also determine conditions of transfer, modification or abolition of use rights. Priorities to use ground water, however, should be kept flexible so as to satisfy present and future requirements such as socio-economic factors.

Economic Measures

1. Economic measures such as fees and waste-disposal charges should be applied in co-ordination and have sufficient impact to constitute an effective incentive to use ground water rationally or be a disincentive to polluting aquifers.
2. The abstraction of ground water could be subjected to differentiated fees in proportion to the volume abstracted, in relation to the available resource or according to the anticipated use of the abstracted ground water, while complying with legal provisions and regulations governing the applied permit system.
3. Costs attributable to pollution should be borne by the polluter whenever the latter can be identified. Serious consideration should be given to all possible economic measures which could have an influence on preventing, mitigating and counteracting damage as well as those bearing on remedying critical situations caused by pollution or over-exploitation of aquifers.

Permit and Penalty System

An appropriate and effective permit and penalty system should be introduced and administered by the competent authorities. The system should promote preventive approaches inducing users to control any activity affecting the quantity and quality of ground water.

Exploration and Abstraction Permits

Permits for ground-water exploration or prospection should be granted by the responsible authority separately from those for ground-water abstraction or use because of the functional difference between the sinking of an exploration well and the large-scale exploitation of aquifers. Exploration permits should have short-term duration.

Abstraction and Recharge Permits

1. Abstraction of water from aquifers and the artificial recharge of ground water should be licensed and controlled by competent authorities according to specific requirements laid down in an appropriate permit system which should be flexible so as to adapt to site-specific conditions. The question of ground-water exploitability should be clarified on a case-by-case basis, taking into account all relevant aspects, including ecological ones. The relevant regulations should establish the extent to which exemptions can be allowed in cases of, for example, ground-water abstraction for households and dystems for draining fields. Where compatible with national legislation, permits for ground-water abstraction and use as well as pollution control should not release the user of ground water from responsibility in case of detrimental effects on ground-water quality and quantity as a result of interventions covered by the permit granted.
2. Specifications of licences should include the purpose, amount, location, duration and technical characteristics of abstraction, as well as the legal status of the ground-water user.
3. Authorization for artificially recharging the aquifer should be granted only if the hydrogeological situation, environmental conditions and the recharge-water quality permit injection, percolation or infiltration of water by artificial means into aquifers for storage and retrieval of good-quality water as well as for restoring over-exploited ground-water resources. For induced recharge from adjacent streams or lakes, appropriate security measures should be applied to forestall accidental pollution.
4. Appropriate measures should be taken to combat saltwater encroachment into coastal aquifers. In such areas special regulations for ground-water abstraction should be enforced to avoid seepage into aquifers owing to over-pumping and the resultant lowering of the ground-water table.

Pollution-Control Permits

1. To prevent ground-water pollution, permits issued to regulate the discharge, disposal and possibly the storage of waste should specifically take into account the vulnerability of the aquifer concerned and the provisions necessary for its protection. These provisions should, in particular apply to production, handling, trading, transporting, storage and use of potentially hazardous substances especially those which are toxic, persistent and bioaccumulative and apply above all to:
   - Effluents and sludges produced by waste-water treatment plants;
   - Domestic-waste disposal sites;
   - Subsurface waste containment by deep-well injection or container storage as *ultima ratio*;
   - Surface storage of wastes potentially hazardous by virtue of their chemical composition.

   In permitting such activities, *inter alia,* the hydrogeological situation of the area should be taken into consideration. On this matter the opinion of qualified specialists should be sought in all above-mentioned cases. Continuous monitoring programmes should be set up both to control water quality in aquifers as well as for checking compliance with permits granted. The specific regulations on nuclear plants and the handling and processing of radioactive substances should include appropriate provisions for the protection of underground waters.
2. Siting of controlled waste disposal should assure that there is no immediate and/or long-term hazard to ground water. Controlled sites should be equipped with protective installations according to the best available technology and monitored by competent authorities. Regulations or guidelines should be drawn up for the site selection of controlled waste-disposal sites, their operation, monitoring, shut-down and eventual rehabilitation, with particular emphasis on ground-water quality protection. Solid or liquid wastes which are toxic, persistent, bioaccumulative or radioactive and which put ground water at risk should be subject to special treatment. Legislation should ban dumping of solid or liquid wastes at unauthorized sites.
3. When applying and operating pollution-control technology for cleaning gases, liquids or solids, during or after treatment processes, the pollutants concentrated in sludges, slurries, gases or solids should not be released uncontrolled into the environment and nor ultimately reach and pollute aquifers. Care should be taken so that pollutants separated from exhaust gas, flue gas and other gaseous emissions do not enter ground water. Measures should be taken to prevent pollution by volatile organic compounds and other aerosols.
4. Pollution control at the source should cover, in particular, those pollutants which are toxic, persistent and bioaccumulative. To this end, regulatory measures and economic incentives should encourage the replacement of chemicals hazardous to ground water by less harmful substances in industrial production processes. Similarly, regulations governing contaminants could be enforced with regard to trade, processing and transport of hazardous substances, with a view to averting or minimizing the risks of leakage into aquifers, and preventing accidental spills of hazardous substances.

5. Application of treated waste water and resulting sludge on land should be subject to licence and/or conform to nationally agreed codes of practice and be restricted to areas where there is no immediate or long-term hazard to ground-water quality. In this respect, particular care should be taken not to overload the self-purification capacity of the soil filter and corresponding natural processes therein. Special attention should be paid to hazardous substances, for example heavy metals.
6. In principle, injection of liquid wastes into the ground should be prohibited. Deep-well injection of liquid wastes of industrial origin and other water of objectionable quality into the ground should be authorized only case-by-case as *ultima ratio,* if the necessary precautions and controls for deep-well disposal can be specified, and if injected wastes cannot harm nearby aquifers. Control methods should include proper siting, design, construction, operation, abandonment and monitoring of deep-well injection. Control measures should be taken to prevent wastes escaping into freshwater aquifers. Continuous hydrogeological surveys should be carried out in the planning stage and during construction. These should serve as a basis for authorization. Permits should specify, *inter alia,* restrictions on the operation programme, emergency procedures in the event of malfunction, as well as monitoring and abandonment procedures for deep wells.

Wells and Boreholes

Drilling and sinking of wells and boreholes should as a rule be carried out by qualified and properly skilled personnel and with appropriate equipment. Prior notice should be given to competent authorities for drilling, sinking, constructing, enlarging, altering, sealing and repairing wells or boreholes and, once work is terminated, a report on the work accomplished should be filed. Provisions might be made for exemptions following precise rules in the case of small and shallow wells. Systematic control should be carried out over the technical status of operation, exploring and monitoring wells, in order to prevent the intrusion into aquifers of polluting substances from the land surface and the mixing of various water layers through drilling of wells.

Monitoring and Control

1. Monitoring programmes for ground-water protection should be set up and applied. These programmes should include monitoring at the source of potential pollution which could pose a serious or chronic threat to an aquifer. There should be regular inspections to ensure compliance with protection requirements imposed. Attention should also be paid to the monitoring of ground-water quality changes brought about by air-borne pollution.
2. Systematic monitoring should be carried out for all aquifers found to be vulnerable to pollution and/or over-use, as well as for those whose particular importance has been recognized for public water supply, mineral water supply and industry.
3. Monitoring and control should be considered a public-service activity. Facilities should be set up for co-ordinating the assessment and availability of monitoring data

and information on aquifers. The resulting collections of data should be related to information on ground-water quantity and quality characteristics of aquifers as well as details of their location, use, and exposure to various impacts from land uses such as agriculture, industry and urban development. Information should be readily available to those interested.

4. The data from monitoring should make it possible to revise periodically plans and forecasts of groundwater use, taking into account actual evolution of aquifers, and to determine measures necessary to ensure the sustainable use of ground-water resources in the long term. Legislative provisions and regulations should, as appropriate, allow for the revision of protection requirements imposed depending on the measures thus determined.
5. Monitoring programmes should be periodically reviewed to ensure that they are achieving their stated aims and that the results have been used effectively.

Land-Use Policies

1. Land-use policies should take duly into account the exigencies of natural recharge and protection of ground water against pollution and over-exploitation. Co-ordination between the various responsible authorities should be promoted. The general application of land-use plans, where appropriate, can be an effective measure in this connection.
2. A co-ordinated approach to ground-water management and land-use control may call for negotiation procedures. Land-use policies should be co-ordinated with other relevant policies of integrated water management such as surface-water and ground-water management policies. Water managers should be involved in land-use planning already at an early stage of development processes. In areas where aquifers are unique, endangered or already impaired, ground-water protection strategies should carry decisive weight in land-use planning and control.
3. Aquifers should be designated critical when already heavily endangered or impaired by pollution or over-use. To avoid further degradation and to make possible their restoration, appropriate measures should be taken which could include changes in land-use patterns and related rights.
4. Sites of waste disposal and places where activities may result in contamination of land should be appropriately restored.
5. Increased attention should be paid to enhancing natural recharge of aquifers.

Protection Zones

1. Where compatible with national legislation, ground-water protection zones should be established over and above the general protection of ground waters through relevant legal provisions, as a preventive measure protecting aquifers around present and future abstraction sites and in recharge areas where aquifers are vulnerable. Compliance with the prescribed restrictions should be strictly controlled.

2. Protection zones could be divided into different classes with differentiated restrictions on land-use and water-use graduated according to environmental considerations and the relative importance of the underlying aquifer. In this respect, particular consideration should be given to the establishment of well-head protection areas. Necessary measures should be taken to minimize the risks of accidental or diffuse pollution in protection zones.
3. Restrictions and/or prohibitions on land-use activities should include mining and industrial processes, manufacturing, intensive farming, including the application of fertilizers and pesticides on agricultural land, transportation, waste disposal and treatment as well as storage of dangerous substances.

Pollution from Agriculture

1. Advice, recommendations, codes of good agricultural practice, legislation, regulations or economic measures should be applied to keep under control the widespread use of fertilizers and any chemicals in agriculture having potential effects on ground water. Policies applied through such measures should take into account the general diffuse nature of such pollution as well as the often considerable time-lag in the transfer of polluting substances to aquifers. These policies should, therefore, promote the application of preventive measures. To that end, all appropriate measures mentioned above should also be implemented to encourage and to promote the rational application of industrial fertilizers, manures, crop-protection products and pesticides. Restrictive measures with penalties for non-compliance should be adopted in particular for the protection of vulnerable aquifers and for protection zones.
2. In order to promote the rational use of agricultural inputs, appropriate measures or a combination of them should be taken, wherever deemed necessary, e.g., the establishment of contractual arrangements between professional agricultural organizations and water authorities; restrictive measures of a legal, regulatory or economic character taking into account socio-economic constraints and environmental conditions prevailing in each country or region.
3. Appropriate measures for controlled use of manure could be, for instance: limiting the operation of intensive large-scale livestock farming to sites where sufficient land is available for correct application and use of manure and slurry; making available large tracts of arable land through co-operative agreement between crop-farmers and livestock farmers, if necessary transporting manure over longer distances to croplands. If in a particular region manure is produced in excess of that needed for plant growth, transport to regions which are deficient should be considered as a solution (manure banks). Conditioning of wastes for stabilization and sale as well as treatment of manure with recuperation of by-products (fertilizers) could also be considered. Manure-application control measures may also include: regulations including dosage and timing; volume increase of manure storage tanks; designing and using better equipment for manure spraying.
4. Strict licensing procedures should be introduced for the manufacture and distribution of crop-protection products and pesticides. Their use and application, however, is

difficult to regulate. Recommendations should nevertheless be made with regard to their dosage, conditions of use, periodicity of application, precautions to be taken, etc. This advice may be included in information campaigns for farmers.

Pollution from Urban and Industrial Sources

1. Measures should be taken to control pollution associated with surface run-off from paved impermeable areas (e.g., streets) and with leakage from industry, transport, sewerage systems and treatment plants. Polluted surface run-off should be properly treated especially where vulnerable aquifers could be affected.
2. Sites of unauthorized waste disposal and other contaminated areas should be identified and adequately restored.
3. Leakage and spillage of contaminants from industries, transport, sewerage systems and waste-water treatment plants should be prevented through appropriate design as well as efficient maintenance and supervision including leakage tests Detection of leakage from pipelines, storage tanks and other industrial facilities should be improved by appropriate inspection procedures.

Control of Mining Activities

1. Dewatering of mines should be kept under control so as to minimize the adverse effects on the water regime, prevent depletion or pollution of nearby aquifers and infiltration of low-quality water. Special attention should be paid to the disposal of mining wastes so that they do not put at risk the quality of ground water. When mines are closed down they should be properly sealed off parallel to the abandoned wells and waste dumps. Vegetation on them should be re-established.
2. In planning and operating *in situ* conversion of coal into natural gas or fluids, strict safety measures should be designed and applied to avoid ground-water pollution induced by processing gases or liquids. This should be done in order to prevent leaching of minerals from reaction zones after mining, or to bar solvents, sulphur and phenol-rich fluids from reaching the ground water. Sites of any kind of waste, in particular tailings ponds for wastes from oil-sand and oil-shale extraction, as well as coal and metal mining should be sealed off against their contaminating ground water.

Research

Research programmes should be intensified in order to improve knowledge of:

(a) hydrodynamics in aquifers;
(b) transportation, fixation and leaching processes of pollutants as well as cumulative phenomena of chemical compounds in the subsoil, in both saturated and unsaturated zones and even in deep-lying aquifers;

(c) appropriate technical measures which prevent, or at least reduce substantially, the transfer of undesirable substances from human activities into ground waters and/or related surface waters;
(d) elaboration of guidelines and technologies with regard to the prevention of ground-water pollution from agriculture;
(e) economical and effective clean-up methods for polluted soils arid aquifers; and
(f) development and calibration of representative ground-water models. International co-operation aiming at the exchange of experience and views and/or joint or co-ordinated research programmes in these fields should be encouraged.

Education and Information

1. Education and information should promote greater awareness of the inherent ground-water problems at all levels, contributing to efficient implementation of the measures taken.
2. Every effort should be made to raise the level of knowledge of the public, in general, and of water users, in particular, as regards the nature, behaviour and vulnerability of ground-water resources. To this end, public information, education and training programmes should be encouraged.
3. Active participation of all parties concerned with the management and use of ground water should be promoted, with a view, *inter alia,* to achieving public acceptance of legal and administrative measures which could restrict the freedom of individual water users, in order to avert possible hazards in the case of misuse. Such knowledge should forestall resistance or outright opposition to the implementation of sound policies for ground-water management.

International Co-Operation

1. Concerted endeavours to strengthen international co-operation for harmonious development, equitable use and joint conservation of ground-water resources located beneath national boundaries should be intensified. To this end, existing or new bilateral or multilateral agreements or other legally binding arrangements should be supplemented, if necessary, or concluded in order to place on a firmer basis co-operative efforts among countries for the protection of those ground-water resources which can be affected by neighbouring countries through exploitation or pollution. In order to implement such co-operation, joint commissions or other intergovernmental bodies should be established. The work of other international organizations, particularly on data harmonization, should be taken into account.
2. Co-operative arrangements could include: data collection, standardization and exchange; establishment of joint inventories; research and training; planning and demand-management; joint control and monitoring of activities with regard to quantitative and qualitative aspects of ground-water protection; elaboration of compatible monitoring methods, standards and permits; establishment of adjacent protection zones; establishment of commonly agreed land-use plans and practices;

monitoring of surface-and ground-water resources' behaviour and interdependence; and the obligation to give notification concerning any activity which might modify the volume and/or the quality of ground water.

5.10. Coping with Changing Water Resources

Although subject to uncertainty, forecasts of climatic change offer a glimpse into possible future water resource impacts and challenges. Predicted impacts vary by region, but include increased temperatures and evaporation rates; higher proportions of winter precipitation arriving as rain, not snow; earlier and more severe summer drought, and decreased water quality. Water shortages, which currently result in substantial economics losses, will be more common in many regions because of these impacts. Such economic losses, which occur across a range of sectors, from agriculture to energy and recreation, have profound effects on local communities. More frequent shortages imply increased costs to society, although adaptation by water users will mitigate some portion of these costs. Water resource users can reduce the negative effects of water shortages through a number of strategies. These include revising water storage and release programs for reservoirs, adopting crops and cropping practices that are robust over a wider spectrum of water availability, expanding and adjusting crop insurance programs (such as the Multi-Peril Crop Insurance program's Prevented Planting Provision), adjusting water prices to encourage conservation and the expansion of water supply infrastructure, and supporting water transfer opportunities. Damage from drought-induced wildfires can be minimized by using long range soil moisture forecasts to pre-position fire suppression resources and in the longer term, by changing land-use regulations to restrict development in areas facing increased fire risk.

The ability to anticipate and efficiently prepare for future water resource management challenges is currently limited, in part, by imprecise regional climate change models and long-term weather forecasts. Uncertainty about future climate conditions makes it more difficult to optimally prepare for and adapt to associated changes in water resource availability and quality. Imagine, for example, trying to prepare optimally for a water shortage when you are uncertain of when it will occur, how severe it will be, or how long it will persist. It may be tempting to make management plans based on the worst-case scenario; however, the opportunity cost of this "safety-first" approach can be high if the worst-case does not occur. Imperfect information ultimately increases the magnitude of economic losses (or reduce the magnitude of any potential economic gains) attributable to water resource changes.

Water resource users can reduce the negative effects of water shortages through a number of strategies. These include revising water storage and release programs for reservoirs, adopting crops and cropping practices that are robust over a wider spectrum of water availability, expanding and adjusting crop insurance programs (such as the Multi-Peril Crop Insurance program's Prevented Planting Provision), adjusting water prices to encourage conservation and the expansion of water supply infrastructure, and supporting water transfer opportunities. Damage from drought-induced wildfires can be minimized by using long range soil moisture forecasts to pre-position fire suppression resources and in the longer term, by changing land-use regulations to restrict development in areas facing increased fire risk. The ability to anticipate and efficiently prepare for future water resource management challenges is currently limited, in part, by imprecise regional climate change models and long-term weather forecasts.

Improvements in climate projections and long-term weather forecasts, such as forecasts based on the El Niño-Southern Oscillation phenomenon (ENSO), offer potential for reducing economic losses (or increasing economic gains) associated with climate change. More specifically, improvements in the ability to detect water shortages farther in advance, to more precisely forecast their location, intensity, and duration, and to use such forecasts to inform management strategies would enhance water users' confidence in regional forecasts, and their ability to efficiently prepare for and adapt to future water resource management challenges.

5.11. Challenges in Water Resources Management and Opportunities

Global climate change will profoundly affect hydrologic systems worldwide. Glacial melting and increasing ocean temperatures lead to sea-level rise. On the continents, the frequency and severity of floods and droughts are expected to increase, while higher temperatures will reduce winter snowpack and hasten spring snowmelt from mountainous areas. Unchecked, groundwater depletion can exacerbate the impacts of these changes; conversely, controlled management of groundwater depletion can contribute to their mitigation.

Historically, society's response to floods and droughts has been to impound surface water in reservoirs, and to release it as needed. However, a dearth of geologically suitable locations for new dams, combined with increased awareness of their ecological consequences, will hinder this response to future hydrologic extremes, even as their frequency and intensity increase. Long-term temperature rises will increase the need to store water for distribution over a longer dry season.

Societies respond to water-resource depletion by shifting management objectives from locating and developing new supplies to augmenting, conserving, and reallocating existing supplies. At the same time, societal objectives are evolving to value water for nontraditional uses, such as maintaining instream flows for aquatic ecosystems. Future groundwater management will have to address these multifaceted challenges. Augmenting supplies can mean improving water quality or increasing water quantity. Depletion due to quality considerations can often be overcome by treatment, whereas large volumetric depletion can only be alleviated by decreasing discharge or increasing recharge. Artificial recharge of storm-flow and treated municipal wastewater, for example, has successfully reversed groundwater declines in some areas. In the future, improved infiltration and recharge technologies will be more widely used to maximize the capture of runoff and treated wastewater. Conserving groundwater by reducing pumpage can be accomplished through administrative, legislative, or management controls, including economic incentives to reduce demand. It is important to target reductions that actually save water. In agricultural areas, for example, improved efficiency is sometimes sought through lining irrigation canals to reduce seepage. But this approach saves no water if the leaky canals are themselves a major source of recharge to the underlying aquifer, as in the North China Plain. If on-farm efficiency gains in saving water are used to irrigate additional land, there will be no overall reduction in water consumption.

Reallocating water resources will play an increasingly important role in groundwater management. Water markets, leasing, trading, and other mechanisms can move limited water from lower to higher productivity sectors, as an alternative to further depletion. Effective reallocation requires rules to ensure fairness and minimize damages. When large-scale groundwater development began, no institutional mechanisms were in place to control the amount of withdrawals. In contrast to large-scale surface-water systems, which are centrally managed, groundwater supplies were mostly "managed" by individual users. Thus, groundwater development has been largely unregulated, even in many water-scarce areas.

Some regions, particularly in semi-arid and arid climates, may follow the lead of Saudi Arabia, which abandoned its goal of grain self-sufficiency through irrigated agriculture when groundwater mining could not be sustained. In other areas, large-scale water transfer projects might maintain activities and populations that depend on or benefit from the depletion of groundwater resources, even at the expense of environmental impacts in the water-exporting basin.

Managers of both surface and groundwater will face new challenges of fulfilling not only the traditional objectives of securing water supplies, but also of improving and protecting ecological health, while facing greater climatic fluctuations and population pressure. To achieve consensus, managers must balance the competing needs of people, industry, agriculture, and the environment. At present, many developed countries that place high value on ecological health of springs, wetlands, and streams have the ability to engineer solutions to help meet these complex challenges. In developing countries, where the livelihoods of millions of poor people may depend on unsustainable groundwater withdrawals, water managers face additional complexities that are not amenable to engineering solutions alone. In the future, the pressure of increasing populations worldwide may foster greater acceptance of groundwater depletion, regardless of a nation's development stage.

In the next few decades, groundwater depletion will likely continue to grow, but at a reduced rate. The change in trend is already in evidence in several depleted aquifers in the western US, and results in large part from positive management actions, but also to some degree from the tendency towards self-limitation of depletion imposed by hydraulic and economic constraints. Land subsidence caused by groundwater exploitation in China has been documented by Xue et al. (2005). The challenge for hydrogeologists is to develop and apply innovative technical approaches, built upon a solid scientific foundation, that credibly inform society of the impacts and alternatives to groundwater depletion.

Relevant Journals

- International Journal of Water Resources Development (Butterworths Scientific Ltd, UK)
- Water International
- Water Resources Bulletin (American Water Works Association)
- Water Resources Research (American Geographical Union)
- Water Resources

- Water Resources Management
- Advances in Water Resources (Elsevier)
- Water and Environment Journal
- International Journal of Water
- J. of Water Resources Planning and Management (ASCE, USA)
- Agricultural Water Management
- J. of Irrigation and Drainage Division, ASCE
- Transactions of the American Society of Agricultural Engineers
- Journal of the Institution of Civil Engineers, London
- Ground Water (American Water Works Association)
- Ground Water Monitoring & Remediation
- J. of Water Chemistry and Technology
- Journal of Hydrology (Elsevier)
- Irrigation and Drainage System
- J. of Applied Irrigation Science (Germany)
- Hydrological Sciences Journal (Blackwell, UK)
- Environmental and Resource Economics

Questions

1. Briefly explain what do you mean by water resources management?
2. Narrate the basic requirements of management of a resource?
3. What are the principal approaches of managing water resource?
4. Discuss the various means/options of demand side management of water?
5. Briefly explain the obstacles & potentials of demand side management in your province/state.
6. Discuss in brief the various options of supply side management of water resource.
7. What is sustainable development? Is sustainable development of water resource possible in your area?
8. Discuss the strategies to achieve sustainability in water resource.
9. Explain in brief, the impact of climate change on water resource.
10. What adaptive alternatives do you suggest in water resource management to cope with climate change?
11. What are the forthcoming challenges the mankind will face with respect to water resources? What preventive/remedial measures do you suggest against the challenges?

REFERENCES

Ali, M. H. & Adham, A. K. M. (2007). Impact of climate change on crop water demand and its implication on water resources planning – Bangladesh perspectives. *J. of Agrometeorology*, *9*(1), 20 – 25.

ASCE Task Committee for Sustainable Criteria, (1998). Sustainable criteria for water resource systems. ASCE Division of Water Resources Planning and Management, ASCE Press, Reston, VA, 253pp.

Griffin, R. W. (1997). Management. 5th Ed., India, A.I.T.B.S.P and D.

Salman, S. M. A. (2007). The Helsinki rules, the UN watercourses convention and the Berlin rules: Perspectives on international water law. *International Journal of Water Resources Development*, *23*, 625–640.

Sauer, T. J., Hatfield, J. L. & Prueger, J. H. (1996). Corn residue age and placement effects on evaporation and thermal regime. *Soil Sci. Soc. Am. J.*, *60*, 1558-1564.

Sivakumar, B. (2013). Planning and management of shared waters: hydropolitics and hydropsychology – two sides of the same coin, International Journal of Water Resources Development, DOI: 10.1080/07900627.2013.841072.

UNESCO (United Nations Educational, Scientific, and Cultural Organization), (2009). Atlas of transboundary aquifers: Global maps, regional cooperation, and local inventories. Paris: International Hydrological Programme.

Waggoner, P. E. (1990). Climate Change and U.S. Water resources. John Wiley, New York.

WCED (World Common on Environment and Development) (1987). Our Common Future. The United Nations, New York.

Chapter 6

GROUNDWATER CONTAMINATION

ABSTRACT

Water supply based on clean groundwater is both economically and environmentally desirable and many countries rely partly or solely on groundwater supply. Groundwater can be polluted by landfills, septic tanks, leaky underground gas tanks, and from overuse of fertilizers and pesticides. If groundwater becomes polluted, it will no longer be safe to drink. A contamination of the resource will pose a threat to human health directly by drinking, and indirectly through food grain and other agricultural products.

6.1. CONCEPTS

In areas where material above the aquifer is permeable, pollutants can readily sink into groundwater supplies. Groundwater can be polluted by landfills, septic tanks, leaky underground gas tanks, and from overuse of fertilizers and pesticides. If groundwater becomes polluted, it will no longer be safe to drink.

More than half of the world's residents depend on groundwater as their primary source of fresh drinking water - either through public water supply systems or private drinking water wells. For many communities, groundwater is the only possible source of fresh water for drinking. Thus, a contamination of the resource will pose a threat to human health directly by drinking, and indirectly through food grain and other agricultural products. The crop plants are also sensitive to some major pollutants. Once contaminated, cleanup of groundwater contamination sites is expensive and slow, and often creates hardships for the persons affected. Groundwater therefore needs to be regulated and protected. It is important for all of us to learn to protect our groundwater because of its importance as a source of water for drinking and irrigation. In order to do this effectively, the threats to groundwater first need to be identified and defined, so that regulatory instruments can be designed to deal with the threats.

6.2. PRINCIPAL THREATS

The principal threats to groundwater resource are:

- Pollution/contamination sources
- Over-exploitation
- Climate change
- Population growth and urbanization (through excessive or over-exploitation, and reducing recharge surface)

Groundwater contamination may be localized or spread over a large area, depending on the nature and source of the pollutant and on the nature of the groundwater system. Identification of threats provides a sound scientific basis for future remedial actions and long-term monitoring.

6.2.1. Sources of Groundwater Pollution

Any substance that is placed or injected in the ground has the potential to affect groundwater quality. Businesses such as dry cleaners, photographers and hair salons serve as examples of potentially hazardous land uses due to the types of chemicals they routinely use. If these businesses operate on individual well and septic service, the chance of groundwater contamination, through an accidental spill or mishandling, is especially high. Other businesses normally considered environmentally sound, such as golf courses, can actually threaten groundwater. These businesses often use relatively large amounts of lawn chemicals. Directly applying these chemicals to the ground presents an uninterrupted opportunity for groundwater contamination.

Substances that can contaminate groundwater can be divided into two basic categories:

1. substances that occur naturally, and
2. substances produced or introduced by man's activities.

Substances that occur naturally include minerals such as iron, calcium, selenium, arsenic, etc. Substances resulting from man's activities include synthetic organic chemicals and hydrocarbons (e.g., solvents, pesticides, petroleum products), landfill leachates (liquids that have dripped through the landfill and carry dissolved substances from the waste materials), containing such substances as heavy metals and organic decomposition products, salt, bacteria, and viruses.

Based on the sources of groundwater contamination, contaminants can be broadly classified as:

- Point source, and
- Non-point source (or diffuse source)

Table 6.1. Groundwater contaminants and pollution sources

Pollution source	Type of contaminant
Agricultural fields	nitrates; ammonium; pesticides
Pesticide Manufacture	Arsenic, phenols, and various halogenated hydrocarbons
Solid Waste Disposal	Heavy metals, ammonium
Sewage Sludge Disposal	Nitrates, various halogenated hydrocarbons, lead, zinc
Gasoline Filling Stations & Garages	Benzene, phenols, and other aromatic hydrocarbons
Metal Industries	Heavy metals, phenols, cyanide, trichloroethylene, tetrachloroethylene, and other halogenated hydrocarbons
Painting and Enamel Works	Metals, some aromatic hydrocarbons, alkylbenzene, tetrachloroethylene, other halogenated hydrocarbons
Timber Industry	Pentachlorophenol, some aromatic hydrocarbons
Dry Cleaning	Trichloroethylene, tetrachloroethylene

Point Source

In this case, the pollutants/contaminates originates from one single point. Examples are septic tank, industrial outlet, dairy farm, landfill/waste disposal site, etc. Some sources of potential groundwater contamination are somewhat easier to identify. They include industrial operations which may use hazardous chemicals, landfills, gasoline filling stations, and other direct sources of contaminates (**Table 6.1**). In most cases, these sources are heavily regulated by the state or federal government. These include regulations affecting landfill, hazardous wastes, and underground storage tanks (removal and construction). In general, the point sources are easier to identify and manage.

Non-Point Source/Diffuse Source

Non-point source or diffuse source are those that lack a well defined single point of origin, such as those created by intensive use of fertilizers, herbicides, and pesticides in agricultural fields. In addition, small point sources, such as numerous domestic septic tanks or small accidental spills from both agricultural and industrial sources, threaten the quality of regional aquifers. A problem of growing concern is the cumulative impact of contamination of a regional aquifer from nonpoint sources.

Pollution from diffuse sources has been a problem for groundwater for some time. Anything soluble which is put on the land has the potential to get into an underlying aquifer, so everybody needs to be careful about what they put on gardens, roads and fields.

Nitrate

Nitrate is one of the most common groundwater pollutants. It can come from widespread sources, like concentrated livestock excretion, the use of nitrogenous fertilizers on farmland and nitrogen-fixing plants, or leaking sewers. More than two thirds of the nitrate in groundwater comes from past and present agriculture, mostly from chemical fertilizers and organic materials. Organic materials, such as manure or treated sewage sludge, can be a valuable source of nutrients and organic matter to soils. If too much is applied, or is applied in the wrong place or at the wrong time, it can get into and harm groundwater. It is estimated that over 10 million tons per year of organic material is spread on the land in the UK. More than 90 per cent of this is animal manure, the rest is treated sewage sludge, green waste compost, paper sludge and organic industrial wastes.

High nitrate levels with inappropriate land use management activity is often a significant cause of diffuse-source pollution. Other major sources of nitrate are leaking sewers, septic tanks, water mains and atmospheric deposition. Atmospheric deposition of nitrogen makes a significant contribution to nitrate inputs to groundwater. Sources of atmospheric deposition of nitrate include transport, power generation and farm animals.

6.2.2. Pesticides and Herbicides

Pesticides are used to control weeds and pests, and can cause diffuse pollution. They can get into groundwater if they are able to leach through the soil. Some pesticides break down slowly, so will stay in groundwater for a long time. Others break down quickly, but some produce more toxic chemicals in the process. Some pesticides have been banned from use but we need to make sure that the replacements are better for the environment. Herbicides can pollute shallow groundwater systems if not managed correctly.

6.2.3. Solvents, Hydrocarbons, and Fuel Additives

Fuel, fuel additives and solvents can contaminate groundwater under cities and industrial areas. Some of these chemicals are quite new, so we do not know how much of a problem they will be. Chlorinated solvents are widely used as degreasers in the metal engineering, electronics and leather industries. They are denser than water, they degrade slowly and are toxic at very low levels so small amounts can pollute large volumes of groundwater. There is widespread contamination possibility in the aquifers below our industrial cities.

Groundwater can be contaminated by hydrocarbons (such as oils and fuels) from installations such as petrol stations that are not properly installed or maintained. Hydrocarbons can disperse in water and be transported over long distances.

6.2.4. Naturally Occurring Substances

Contamination also results from an overabundance of naturally occurring iron, sulphides, manganese, and substances such as arsenic. Excess iron and manganese are the most common natural contaminants. As water flows through the ground, the chemistry changes as elements are released from the rocks. In many aquifers around the world, iron, manganese, arsenic, fluoride and radon are found at relatively high concentrations. Another form of contamination results from radioactive decay of uranium in bedrock, which creates the radioactive gas radon. Methane and other gases sometimes cause problems. Dissolved radon can be released as a gas and accumulate in confined spaces. Seawater can also seep into groundwater and is a common problem in coastal areas. It is referred to as "saltwater intrusion."

The high risk of contamination by these compounds is because they are very mobile (as leachate) in groundwater environments. The improper management of hazardous substances has contributed directly to localized and regional ground water contamination throughout the world. It only takes a small amount of contamination to impact an aquifer and raise health concerns.

6.2.5. Others

Salinity

There is a natural boundary where fresh and saline groundwater meet in coastal aquifers. When fresh water is pumped out, this boundary moves and seawater is sucked into the aquifer. This can also happen inland, with groundwater abstraction causing saltwater to be drawn up from deep water aquifers. If too much water is taken from a coastal aquifer, it can eventually become too salty to drink.

Microbes

Potential sources of microbiological contaminants are sewage and effluent.

Mining

Metal and coal mines can have a huge impact on groundwater systems. While mining is happening groundwater is usually pumped out in large quantities. When mining and pumping stops the rising groundwater can become contaminated. The mine structures can also change groundwater flow permanently.

The main pollutants from mining are iron, zinc, lead, cadmium and acidic water. These substances can leach down into groundwater from spoil heaps, or can contaminate the upper part of the aquifer as groundwater levels (previously artificially lowered by pumping) begin to rise. The impacts can remain for hundreds of years. Metal mining has caused high concentrations of heavy metal pollutants in groundwater across Wales, the south west and northern England.

Landfills

Landfills are another major source of contamination. Landfills are the places that our garbage is taken to be buried. Landfills are supposed to have a protective bottom layer to prevent contaminants from getting into the water. However, if there is no layer or it is cracked, contaminants from the landfill (car battery acid, paint, household cleaners, etc.) can make their way down into the groundwater.

Many different types of waste go to landfill, so there are lots of potential pollutants. They are only a problem if the landfill has not been lined properly, or is badly maintained.

Road-Salts

The widespread use of road salts and chemicals is another source of potential groundwater contamination. Road salts are used in the winter-time to melt ice on roads to keep cars from sliding around. When the ice melts, the salt gets washed off the roads and eventually ends up in the water.

Contaminated Land

Industry can leave behind land contaminated by chemicals or radiological material that can leach through soils and rock and pollute groundwater.

Subsurface Constructions

Large subsurface constructions such as train tunnels and buildings often have a large impact on groundwater flow, especially in the construction phase, where groundwater lowering may change flow conditions considerably.

Contaminants from Other Parts of Hydrologic Cycle

We have to remember that since groundwater is part of the hydrologic cycle, contaminants in other parts of the cycle, such as the atmosphere or bodies of surface water, can eventually be transferred into our groundwater supplies.

6.2.6. Well Contamination

A well can easily be contaminated if it is not properly constructed or if toxic materials are released into the well. Toxic material spilled or dumped near a well can leach into the aquifer and contaminate the groundwater drawn from that well. Contaminated wells used for drinking water are especially dangerous. Wells can be tested to see what chemicals, pathogens and other contaminants may be in the well. A well connects the land surface to an aquifer. This connection can be exploited by contaminants released near the well. The contaminants can travel directly from the surface into the aquifer via the same bore hole used to draw water from the aquifer. Wells can also connect aquifers across aquitards, allowing contaminants from one aquifer to reach another. Seals on unused, orphaned, and improperly abandoned wells can deteriorate over time, creating a connection from the surface to the aquifer. Unused wells can become an important contaminant pathway.

6.2.7. Pollution Potential in Fractured Limestone

In most basic groundwater textbooks, you'll find hydrogeology discussed in terms of porous media, with aquifers composed of well-sorted sand, and groundwater moving through pores between sand grains. In the dolomite terrain (e.g., of northeastern Wisconsin) or cracking clay soil (e.g., in Australia), however, the story is much different. In such zone, the fracture or crack is exposed at or near the land surface. The vertical and horizontal fractures in the underlying soils are typical features. Rain or snow falling on this landscape enters the groundwater system through an interconnected network of vertical and horizontal bedrock fractures. In such systems, rapid groundwater movement and minimal contaminant attenuation are common, and so the land-use practices in the areas where the water originates (often called the capture zones or contributing areas) highly influence the quality of groundwater produced by local wells. Determining these capture zones and understanding how groundwater moves from recharge to local wells are critical to protect groundwater in fractured-rock terrains. Undertaking wellhead-protection studies in fractured-rock settings is a challenging endeavor.

6.3. Pathways for Contaminant Transport

As water percolates downward into the ground, pollutants, present at or below the ground surface may become dissolved. Once dissolved, the contamination may easily seep into the aquifer. Ground water moves from areas where the water table is high to where the water table is low. When water is being pumped from a well, the water table in the immediate vicinity of the well is lowered, thereby increasing ground water flow in the direction of the well. The resulting 'Zone of Contribution' represents the area within the aquifer, which contributes recharge to the well. As such, contamination which enters the Zone of Contribution will eventually move towards and pumping well.

There are hundreds of chemicals that could get into groundwater. We look for substances that are not a problem now, but that we think might be in the future. Pharmaceuticals have been found in groundwater in many countries, usually near landfills that have been used to dispose of hospital or pharmaceutical industry waste.

In a sand and gravel aquifer, contaminants move by diffuse flow. The gasoline from the underground storage tank spreads very slowly and contaminates the lake. Contaminants in a fractured rock aquifer move quickly through the fissures in the rock by conduit flow. This type of flow is less predictable than diffuse flow. When pumping occurred, the contaminant plume moved from one well to another and to the stream.

Understanding pathways for aqueous transport of contaminants is necessary for determining the location and mass of contaminants at a given time, predicting their migration throughout the site's hydrogeologic system, and estimating if and when there might be impacts on regional groundwater. To better characterize potential pathways for contaminant transport, monitoring wells must intercept the contaminant pathways. Toward developing a monitoring program, understanding of pathways is essential for:

- Planning the locations of wells to sample the alluvial groundwater, perched-intermediate groundwater, and the regional aquifer so that the wells are most likely to intercept a contaminant plume;
- Determining the well-sampling frequency and types of analyses needed; and
- Providing a rationale or model for interpreting the sample results.

Geochemistry is central to understand the extent to which contaminants move with groundwater; it is a tool for better understanding hydrogeologic pathways; and it is essential for determining the degree to which monitoring data are representative of actual groundwater. The specific need is to understand how contaminant migration caused by groundwater is affected by geologic or anthropogenic media that are encountered along the groundwater's flow-path.

6.4. Arsenic Contamination – A Major Threat for Groundwater

The occurrence of metallic and non-metallic elements (e.g., iron, manganese, arsenic, barium, etc.) in the groundwater is considered to be natural. Arsenic in groundwater above the maximum allowable concentration level (MCL) has been found more widespread and prevalent throughout the world than previously thought.

Arsenic occurs naturally in the environment and is found in groundwater. Arsenic can occur as a free element, but it most commonly found as arsenopyrite (FeSAs). The concentrations of naturally occurring arsenic in groundwater can vary widely. Arsenic concentrations average 580 micrograms per liter (µg/l) in the earth's crust, 0.01 µg/l in streams. Complex chemical reactions and biological activity can alter immobile forms of arsenic and cause them to dissolve into groundwater. Arsenic is principally used as a poison and in semiconductors. The current MCL for arsenic is 10 µg/l as suggested by WHO and in many countries. Acute exposure to arsenic above the MCL can produce disease in the respiratory, gastrointestinal, cardiovascular, nervous, and hematopoietic (blood cell production) systems. Arsenic occurs naturally in groundwater as a result of geothermal activity and dissolution of iron oxides and sulfide minerals. The adsorption of arsenic to iron oxydroxides is strongly influenced by pH, oxidation/reduction (redox) potential, and the presence of competing ions (Stollenwerk, 2003). Chemical reactions involving iron oxides and sulfide minerals appear to influence the mobility of arsenic. An increase in pH can cause arsenic to dissolve into groundwater. Changes in redox potential can change the number of available adsorption sites, reducing or increasing the amount of arsenic in groundwater. Microbial action can promote the release of arsenic into groundwater (Cummings et al., 1999; Roller et al., 2003). Conversion of sulfides to sulfates by introducing oxygen into an aquifer system can release of arsenic into groundwater.

Relevant Journals

Journal of Water Resource and Protection
Air, Soil and Water Research:
Applied Geochemistry
Applied Water Science
Asian J. of Water Environmental Pollution
Australian J. Agril Engg
Australian Journal of Soil Research
Engineering Geology
Agricultural Water Management

Questions

1. What are the principal threats to contaminate water resources?
2. Discuss the natural and man-made activities that contaminate water resources.

3. Briefly narrate the contaminate sources from agricultural field and industrial sites.
4. How the wells can be contaminated?
5. Discuss the pollution potential in Fractured Limestone
6. Discuss the Pathways for contaminant transport to groundwater
7. Briefly discuss the threat of Arsenic contamination to groundwater

REFERENCES

Cummings, D. E., Caccavo, Jr. F., Fendorf, S. & Rosensweig, R. F. (1999). Arsenic mobilization by the dissimilatoryFe(III)-reducing bacterium Shewanella algaBrY. *Environ. Sci. Technol.,33*, pp. 723–729.

Roller, J., Schreiber, M., Widdowson, M. & Johnson, J. (2003). Arsenic release due to dissimilatory reduction of iron oxides in petroleum contaminated aquifers. Proc. Of the 'Virginia Water Research Symposium 2003' Water resources management for the Commonwealth, Oct. 7-10, 2003, p. 129-132.

Stollenwerk, K. (2003). Geochemical Processes Controlling Transport of Arsenic in Groundwater: A Review of Adsorption. *Arsenic in Ground Water*, 2003, 67-100, DOI: 10.1007/0-306-47956-7_3.

Chapter 7

PROTECTION OF WATER RESOURCES

ABSTRACT

Groundwater pollution or degradation is the main threat to the achievement of sustainable water resources management. Clean-up measures always have a high economic cost and are often technically problematic. Thus proactive campaigns and practical actions to protect the natural quality of groundwater are widely required.

7.1. INTRODUCTION

In most part of the world, groundwater supplies up to 90% of the drinking water that we get through our taps. It is crucial that we look after this source and ensure that our water is completely safe to drink. Groundwater protection is vital to states/provinces where majority of the people draw their drinking water from underground aquifers. Identification and control of contamination sources are essential for groundwater protection.

Most of the problems affecting groundwater are associated with human development. The most common problem posing a threat to groundwater supply is degradation of the water quality due to surface-level pollution or saline intrusion. Groundwater pollution or degradation is the main threat to the achievement of sustainable water resources management. Worldwide, aquifers (geological formations containing usable groundwater resources) are experiencing an increasing threat of pollution from urbanization, industrial development, agricultural activities and mining enterprises. In some cases it may take many years or decades before the impact of a pollution episode by a persistent contaminant becomes fully apparent in groundwater supplies abstracted from deeper wells. This can lead to complacency over the pollution threat. But the real implication is that once groundwater quality has become obviously polluted, large volumes of aquifer are usually involved. Thus clean-up measures nearly always have a high economic cost and are often technically problematic. Thus proactive campaigns and practical actions to protect the natural (generally excellent) quality of groundwater are widely required, and can be justified on both broad environmental-sustainability and narrower economic-benefit criteria.

Controlling a source of aqueous waste could involve treating that waste to remove contaminants or reducing or stopping the discharges. Controlling solid waste could involve ensuring that it is emplaced in such a way that it cannot release contaminants or, if necessary, recovering the disposed waste, repackaging it, and possibly shipping it offsite. We should regulate onsite wastewater disposal in a manner that is protective of the groundwater and in such a way as to minimize the possibility of endangering the health and welfare of the public. The overall aim of the groundwater protection scheme is to preserve the quality of groundwater, particularly for drinking and irrigation water purposes, and maintaining/sustaining the groundwater reserve for the benefit of present and future generations. Once the groundwater protection zone is defined for an area, the hazards posed by selected human activities can be evaluated to assess the appropriate risk management measures, or Groundwater Protection Responses, for those activities.

Datasets of vulnerability, aquifer potential, and source protection areas can be generated by the GSI using field mapping in combination with readily available existing information and a limited amount of targeted drilling and testing. The vulnerability and aquifer datasets cover the entire land surface of a given area, while the source protection area datasets are specific to the catchments of selected groundwater supply sources. We should work towards a groundwater protection strategy suitable to Nation's physical and socio-economic environment.

Groundwater protection program should be made up of the following elements:

- Identification
- Prevention/control
- Monitoring
- Restoration, and
- Communication

7.2. DELINEATING PROTECTION ZONES

Before a plan or program can be developed to protect ground water, it is important to identify existing or potential threats to the ground water. This will generally mean conducting an inventory to learn the location of facilities using, manufacturing, or storing materials that have the potential to pollute ground water. Groundwater protection zones should be developed based on local conditions and different concept and principles. Criteria commonly used for zoning includes the following (Chave et al., 2006):

Distance: The measurement of the distance from the abstraction point to the point of concern such as a discharge of effluent or the establishment of a development site.

Drawdown: The extent to which pumping lowers the water table of an unconfined aquifer. This is effectively the zone of influence or cone of depression.

Time of travel: The maximum time it takes for a contaminant to reach the abstraction point.

Assimilative capacity: The degree to which attenuation may occur in the subsurface to reduce the concentration of contaminants.

Flow boundaries: Demarcation of recharge areas or other hydrological features which control groundwater flow.

Table 7.1. Tracers for groundwater study

Tracer type	Examples
Natural environmental isotopes (stable/unstable)	^{13}C, ^{14}C, ^{2}H, ^{3}H, ^{18}O, ^{3}He, ^{36}Cl, ^{34}S, ^{4}He, ^{39}Ar, ^{85}Kr, ^{15}N, ^{234}U
Radioactive tracers	^{60}Co, ^{3}H, ^{24}Na, ^{51}Cr, ^{82}Br, ^{131}J,
Fluorescent dyes	Rhodamine B, Amidrhodamin G, Uranine

Approaches using such criteria range from relatively simple methods based on fixed distances, through more complex methods based on travel times and aquifer vulnerability, to sophisticated modelling approaches using log reduction models and contaminant kinetics. Uncertainty of the underlying assessment of contamination probability is reduced with increasing complexity. Use of tracer test in determining travel time and flow direction of the contaminant is highly recommended. Commonly used tracers are described in **Table 7.1**.

The program consists of four interconnecting elements: (1) preventing pollution of the groundwater, (2) monitoring the effectiveness of engineered and administrative controls at operating facilities and groundwater treatment systems, (3) restoring the environment by cleaning up contaminated soil and groundwater, and (4) communicating with stakeholders on groundwater protection issues.

7.3. Measures for Controlling Quality Degradation

A strategy for developing a protection plan for the aquifer should be based on the assessment of risk for contamination of the aquifer and the relative value of the water to the society. The strategy may include:

- Source-water protection
- Mitigation measures
- Laws and regulations
- Community involvement
- Practical method

Source-Water Protection

Contaminated groundwater is very difficult and expensive to clean up. Solutions can be found after groundwater has been contaminated but this isn't always easy. The best thing is to adopt pollution prevention and conservation practices in order to protect important groundwater supplies from being contaminated or depleted in the first place. 'Source Water Protection' is a process that enables citizens to protect local groundwater supplies which serve as a source for drinking water.

For source-water protection, we have to define Source Protection Zones (SPZs) for groundwater sources such as wells, boreholes and springs used for public drinking water supply. These zones show the risk of contamination from any activities that might cause pollution in the area. The closer the activity, the greater the risk. The maps of SPZ show three main zones (inner, outer and total catchment) and a fourth zone of special interest, which we occasionally apply, to a groundwater source.

We can use the zones in conjunction with our Groundwater Protection Policy to set up pollution prevention measures in areas which are at a higher risk, and to monitor the activities of potential polluters nearby.

Use of regulations, land acquisition, and education programs can play a key role in protecting groundwater. Examples of land use control activities include the following:

- Land use plans which take into account groundwater vulnerability;
- Zoning ordinance and site plan review standards related to aboveground secondary containment, interior floor drains, and other topics;
- Purchase of land and/or conservation easements to provide a wellhead protection buffer around municipal wellfields; and
- Public education through public meetings, school-based classroom programs, library displays, cable television videos, public information flyers, and municipal newsletters.

Protection of groundwater resources requires efforts on several fronts, including the need for regional planning, land planning for individual sites, and technological advances that may offer alternative solutions. Regional planning must be based on the entire Watershed; it will do little good for one community to implement solutions to its problems only to find that neighboring communities do not. Groundwater has no respect for community boundaries.

From a land planning perspective, simply requiring larger lots does little or nothing to enhance groundwater quality. One of the few readily available solutions to polluted wells or failed septic systems is to obtain public water and sewer. With the larger lots sizes and frontages prevalent in many of the communities within the watershed, the costs to provide water and sewer services to homes are likely to be exorbitant. On the other hand, where lot frontages are lower, so too will be the cost to provide public utilities.

Several regulations are in place by the counties within the watershed which target the protection of water resources. Septic system regulation, for example, is the responsibility of the various Health Departments in relation to permitting, placement, and enforcement. The county health departments are also responsible for the inspection of septic systems prior to the sale of a parcel of land. If the system fails the counties' tests, the system must be upgraded or maintenance must be completed before a permit will be issued to the new property owner.

Septic System Maintenance

It is generally recommended that septic tanks be pumped out or the sludge and scum layers be measured at least every three years so that solids don't wash out into the soil treatment system. Solids can clog the soil and limit its ability to properly treat the septic-tank effluent. A local government may choose to impose septic system maintenance requirements on individual developments.

However, implementing this recommendation is sometimes difficult without adequate cooperation between the community and county health departments. One solution may be to adopt a local ordinance that establishes "septic system maintenance districts," where higher concentrations of septic systems are present. Within these districts, property owners could be required to submit evidence of inspection or maintenance of septic systems at periodic intervals. This may be particularly effective for any approved Open Space Developments which have homeowner associations.

Secondary Containment

A common groundwater protection method for identified potential sources of groundwater contamination (such as above ground fuel storage tanks and facilities) is a requirement for secondary containment. A variety of containment methods are possible, but the most common is the construction of "traps" for runoff and spillage areas where possible contaminants are contained within walls or other structures and any runoff captured and contained within the structure.

Local units of government through the site plan review process should be made aware of the locations for possible contamination and the measures planned by the operator to reduce the risks associated with those materials. This can be done as part of the site plan review process for potential point sources of contamination, such as industrial uses involving chemicals or hazardous materials. Each site plan should contain a storm-water management plan that details the impact of proposed land use on water quantity and quality, both on-site and within the watershed. This may be implemented by adding a requirement for site plans for submission of information regarding potential hazardous materials, and measures to be taken to protect drainage ways, water bodies, or other areas from accidental spills.

Another provision which can be made part of the site plan review process, as well as for discretionary zoning approvals, such as planned unit developments and special land uses, is a requirement for monitoring wells for specific land uses with potential to affect groundwater resources. Monitoring wells have long been required for certain uses, such as landfills. Increasingly, communities are requiring them for other uses, such as golf courses, sand and gravel mining operations, higher density residential developments in identified groundwater vulnerability areas, and others.

On-Site Community Treatment Systems

In cases where soil conditions are inadequate to support a conventional onsite wastewater disposal system, we have to consider the site for an approved alternative system. The expense involved in resolving groundwater issues for a single site makes some solutions financially difficult. One area-wide solution intended for limited use is a package treatment system which serves smaller areas. Although a single, small development project may not be able to afford the installation and operation of a compact treatment system, several combined projects may join forces to implement an effective waste treatment system.

A number of management and financial issues would obviously need to be resolved before such a system is implemented. Administering the system will likely be the responsibility of individual property owners formed into an association or authority. Questions of who will pay for the initial acquisition and installation of the system as well as maintenance responsibilities will need to be addressed. Issues of liabilities and other legal problems must also be examined. Generally, engineering expertise will be needed to conduct routine repairs and inspections, and replace system components when needed.

Wellhead Protection

The purpose of wellhead protection programs is to protect public water supplies taken from groundwater from potential sources of contamination. Protection is provided by identifying the area supplying groundwater to the community's wells, identifying potential sources of contamination within that area, and developing methods to cooperatively manage the area and minimize the potential threat to groundwater.

Wellhead protection programs must address the following elements:

- Establishment of roles and duties for communities and property owners within the wellhead protection area.
- A description of the wellhead protection area.
- Identification of potential sources of contamination within the wellhead protection area.
- Procedures to manage the protection area and minimize threats to the water supplies.
- Plans for water supply emergencies.
- Procedures for the development of new well sites.
- Public participation methods.

A wellhead protection area is defined as the surface and subsurface area surrounding a water-well *or* well-field, supplying a public water system, through which contaminants are reasonably likely to move toward and reach such water-well or well-field. Zoning and land use measures to protect wellhead areas are generally similar to those that protect open spaces, including purchase of lands, conservation easements, and other similar measures. In many States or Provins, grants are available from the 'Department of Environmental Quality' to assist communities in developing wellhead protection programs and legislation.

7.4. Groundwater Monitoring

Elements of the groundwater monitoring program include installing monitoring wells; planning and scheduling; developing and following quality assurance procedures; collecting and analyzing samples; verifying, validating, and interpreting data; and reporting. Monitoring wells are used to evaluate progress in restoring groundwater quality, to comply with regulatory permit requirements, to monitor active research, and to assess the quality of groundwater that enters and exits the site.

7.5. Laws and Regulations

To provide for the sustainable use of groundwater, governments should protect and manage both fresh and brackish groundwater resources. The development and implementation of policy on groundwater protection may be based on an analysis of threats to groundwater and the development of instruments to address the threats.

In the absence of regulation of groundwater (both abstractions and discharges), with increasing development, the consequences sooner or later will be:

(i) Degradation of the fresh groundwater resources, with more fresh water having to be imported or manufactured, at much higher cost, as natural reserves decline.
(ii) Health risks to consumers, from, for example, unregulated discharges from septic tanks.
(iii) Increased costs for treatment of drinking water to potable standard, if the groundwater is contaminated with hydrocarbons, industrial solvents, or pesticides, for example.
(iv) Increasing dependence on imported fossil fuels to support energy-hungry processes like reverse osmosis.
(v) Loss of tourist revenue if water supply becomes unsafe or unreliable, or if the environment deteriorates.

We also have to adopt regulations and do outreach to carry out those mitigation measures. The measures should be designed to prevent continued movement to groundwater in contaminated areas and to prevent problems before they occur in other areas. Laws and Regulations are a better way to protect groundwater. Groundwater regulations are restrictions to protect groundwater from contamination resulting from use of pesticides in sensitive areas. Background on the development of the new legislation should be directive on protection of groundwater against pollution and deterioration. Different countries have formulated legislative framework for groundwater protection (e.g., European Commission (EC), USA) and gave directive guidelines to the member states.

Sample legislative guidelines of European Commission Groundwater Environment are (EC, 2006):

- Groundwater bodies within River Basin Districts to be designated and reported to the European Commission by Member States.
- Establish registers of protected areas within each river basin districts for those groundwater areas or habitats and species directly dependent on water.
- Establish groundwater monitoring networks based on the results of the classification analysis.
- Set up a river basin management plan (RBMP) for each river basin district which must include a summary of pressures and impacts of human activity on groundwater status, a presentation in map form of monitoring results, a summary of the economic analysis of water use, a summary of protection programmes, control or remediation measures, etc.
- Take into account by 2010 the principle of recovery of costs for water services, including environmental and resource costs in accordance with the polluter pays principle.
- Establish by the end of 2009 a programme of measures for achieving WFD environmental objectives (e.g., abstraction control, prevent or control pollution measures) that would be operational by the end of 2012. Basic measures include, in

particular, controls of groundwater extraction, controls (with prior authorisation) of artificial recharge or augmentation of groundwater bodies (providing that it does not compromise the achievement of environmental objectives). Point source discharges and diffuse sources liable to cause pollution are also to be regulated under the basic measures.

7.6. Regulatory Instruments

Regulatory instruments can take the form of primary legislation (such as an Act of Parliament), secondary legislation (such as regulations published by a Government minister with the backing of enabling legislation), less formal codes of good practice (often aimed at a particular industry sector), and sometimes just public information campaigns.

Regulatory instruments may be as follows:

1. *Licensing of abstractions*
   Licensing of groundwater abstraction allows the quantities, locations and timing of abstraction from an aquifer unit to be controlled and limited to sustainable levels, to avoid long-term degradation of the resource. Exemptions can be built into the licensing regime, so that for example, abstraction for domesticuse below a certain threshold daily quantity does not require a license. This helps regulatory bodies with limited resources to concentrate on the larger abstractions, and makes abstraction licensing more politically acceptable. Desirable features of an abstraction licensing regime include: time-limiting licenses, so that they are not irreversible; some form of license trading, to enable unused licenses to be reallocated and to encourage higher value uses of water; and the ability to add conditions to a license.
2. *Charging for abstraction*
   Abstraction charging (in conjunction with abstraction licensing) is an economic instrument commonly employed to tackle the threat of over-abstraction, as it provides an economic incentive to reduce abstraction, especially if payments are imposed on each unit volume of groundwater abstracted. Often the charge is levied only on abstractions above a defined minimum volume, and then there are bands of charges for increasing volumes. The charge may also vary with the season, and the quality of water abstracted. For example, abstraction from brackish sources can be expected to be at a lower price. Similarly, the use to which abstracted water is put can affect the charge, so that low consumptiveuses are charged at a lower rate. Flexibility can also be retained to exempt certain categories of abstractor.
3. *Licensing of well drillers*
   This is a relatively straightforward administrative regime to put in place, and consists of licensing professionals to carry on activities of groundwater exploration, drilling and well construction. Anyone wishing to drill and equip a well must use a licensed well driller. This ensures that wells are drilled and equipped to the required standards, by people who have been properly trained and are aware of the risks of groundwater contamination. The aim is to avoid contamination of aquifers via poorly-constructed wells. The system also enables information and data on

groundwater to be collected, as well drillers are usually required to submit records of all wells drilled, whether or not they go on to be used for abstraction or disposal.

4. *Effluent discharge consents*
   Discharge consents are an instrument to control the disposal of liquid effluents to land, surface water and groundwater. They regulate the volume and toxicity of wastewater or effluent which it is permissible to discharge in a certain location, and usually control the method of discharge as well. Without them, uncontrolled discharges to groundwater can easily lead to widespread pollution. Once groundwater has been polluted, it is very difficult to clean up. Discharge consents are sometimes made tradable, but this is not always desirable. It can lead to ongoing pollution, for example if a powerful industry buys up unused consents to enable it to carry on with unabated discharges.
5. *Effluent discharge fees*
   These are applied in conjunction with discharge consents, and are designed to increase the cost payable by a polluter for using the natural or man-made environment as a discharge outlet for wastes. The fees usually vary depending on the pollution potential of the effluent, and the location of the discharge (inother words, the sensitivity of the environment receiving the effluent). This is an economic instrument, encouraging a reduction in the use of toxic substances in production processes, or stimulating the pre-treatment of wastewater. The level of fees can be raised until the volume and toxicity of effluents are reduced to an environmentally acceptable level, if that exists.
6. *Groundwater protection zones*
   Groundwater Protection Zones (GPZs) are designed to protect sources of groundwater from contamination, particularly sources of fresh groundwater for public water supply. The regulatory body is given statutory authority to demarcate various zones around a groundwater source to control activities within the zones which could contaminate the source. The zones are usually based on criteria such as the total catchment area of the source, and/or the time of travel for pollutants (Foster et al.2002). The zone immediately around the source receives the greatest protection, with progressively less strict controls on activities in concentric zones moving out from the source. GPZs are an important part of the land use planning process.
7. *Land use planning*
   The objective of land use planning is to secure a spatial pattern of land use development that best serves the cultural, social, environmental and economic interests of the nation, while taking into account the views of the people who live and work in the defined area. Groundwater often suffers from being out of sight and therefore out of mind, and there is a great need to elevate groundwater to the status of a strategic national resource. The land use planning system in general, in conjunction with GPZs, has a very important role in pre-empting threats to groundwater by not allowing certain activities to take place in certain locations, or at least to control the way in which the activities take place. This does not mean that reasonable development proposals should not be allowed to happen, but only that groundwater should be taken into account when the planning decision is made.

8. *Building regulations*
   Building Regulations enable Building Control Officers to monitor the standards of building works with reference to detailed obligatory specifications and minimum standards published in Building Codes, usually controlled through permits. From the point of view of groundwater, the idea is to ensure minimum standards of materials and construction for items such as septic tanks, disposal wells, drainage wells, grease traps and oil interceptors, so that the contamination threat to groundwater is minimized. This is a powerful instrument (if properly enforced), and can make a significant contribution to groundwater protection.
9. *Solid waste disposal controls*
   Leachates from solid waste transfer, disposal and landfill sites can be a serious threat to groundwater quality. Solid waste disposal controls lay down where solid waste and sludge may be disposed of, and what type of waste is permitted to go to landfills and rubbish tips. Techniques include the separation of types of sludge and solid waste so that the appropriate types of waste go for the appropriate type of disposal, whether incineration, landfill, land-spreading or recycling. It is important to give strong backing to the implementation of solid waste management plans, with a cradle-to-grave approach to the management and disposal of solid waste.
10. *Controls on the use of biocides and fertilizers*
    The diffuse application of chemical and microbiological substances to land in agriculture, horticulture, and public amenity for controlling pests or providing nutrients can have a significant long -term impact on groundwater quality. This instrument involves setting up product controls, so that the use of certain substances is banned or at least strictly controlled, especially in areas of groundwater vulnerability. The controls can range from banning the importation to the country (or manufacture within the country) of certain substances, to the publication of good practice guidelines for the responsible use of the substances.
11. *Prosecution of pollution offences*
    Under the environmental law of most jurisdictions it is an offence to cause pollutants or contaminants to enter the environment (including groundwater). This is intended to retard people from polluting the environment, and to ensure that the polluter pays if an offence is committed. Unfortunately, these laws are not always adequately enforced, and major pollution incidents sometimes go unprosecuted. This is often due to difficulties with the level of proof required to achieve a conviction. Consideration may also need to be given to increasing the penalties applicable to pollution offences to increase the deterrence.
12. *Contamination clean-up orders*
    Where contaminated land is identified as being an on-going source of pollution for groundwater, it may be technically feasible and cost-effective to remediate the land, to reduce the detrimental effect on groundwater resources. Many jurisdictions have established procedures for identifying areas of contaminated land, and a process for attributing responsibility for the contamination. It is important to realize that prosecution of polluters is not always possible, sometimes because the contamination is historic, and was not caused by the current site owner. Even if successful, prosecution of polluters does not always lead to the site being cleaned up. Clean-up orders are therefore a useful instrument, especially for historic contamination.

13. *Environmental impact assessments*
Environmental impact assessment (EIA) is a key instrument, feeding into the planning process, for the control of larger projects which could potentially have a detrimental impact on groundwater. They consist of a statutory iterative process by which planning decision-makers can assess the environmental effects of certain categories of developments. They also subject the environmental effects of development projects to expert and public scrutiny in a methodical and transparent manner. A key component of any EIA is the Environmental Statement, which is usually prepared by the developers- consultants, setting out the environmental impact of the project and proposed methods of mitigation, as well as information provided by statutory consultees.
14. *Statutory consultation*
The principal focus of this instrument is on ensuring that there are strong statutory links between the various bodies whose activities might have an impact on groundwater. Whereas an environmental regulator might have control of several key instruments (such as abstraction licensing/charging, discharge consents/fees, licensing of well drillers, etc.), some instruments which are vital to groundwater protection may lie outside the formal powers of the environmental regulator (such as building regulations and product controls). These are controlled by other bodies, such as a Ministry of Works. It is important to ensure that the right bodies are statutory consultees in the processes which impact on groundwater protection (for example, through the land use planning process and the EIA regime).
15. *Environmental education*
This instrument, which is a social process aimed at strengthening a new culture of water, can significantly reduce quantitative and qualitative stress on groundwater. Public education can encourage: leakage reduction and the sensible pricing of water by water utilities; the better treatment and recycling of wastewater; water-efficient techniques of irrigation; the provision of secure supplies of fresh water and decent sanitation in housing areas; and the prudent use of water in all sectors of society. This instrument takes a very wide variety of forms, such as: teaching materials for schools; posters, leaflets, books, audio-visual materials; coverage in the press, radio and television; and the dissemination of user-friendly Codes of Practice designed for specific industrial and commercial sectors (which can be an effective preventative means of achieving groundwater protection objectives without recourse to the enforcement of regulations through court prosecution).

7.7. COMMUNITY INVOLVEMENT

Community involvement can make a program sustainable. How your community conducts the inventory of contamination source will depend largely on the resources available, particularly the number of people available to do the work and funds. The inventory of existing or potential threats to the community's groundwater may be quite long, and it is unlikely that your community will have the resources to address all of these threats. How do community officials decide which threats are the most serious or set priorities? One way is to assess these threats on the basis of their relative risks to the community's ground water. This requires determining which of the specific pollutants are most likely to be released and reach the groundwater in concentrations high enough to pose health risks. In addition to having an incentive to protect its ground water, your community has a number of powers that can be used for that purpose. These include implementing zoning decisions; developing land-use plans; overseeing building and fire codes; implementing health requirements; supplying water, sewer, and waste disposal services; and using their police powers to enforce regulations and ordinances. A few communities have begun developing their own groundwater protection programs using a variety of management tools based on these powers.

Groundwater Monitoring: This should be done to assess the quality of local aquifers by sampling public and private wells for selected contaminants.

Household Hazardous Waste Collection: To alleviate the threat to groundwater from the disposal in regular trash pick-ups, sewers, or septic systems of household products that contain hazardous substances or other materials that can be harmful to groundwater, such as paints, solvents, or pesticides.

Water Conservation: To reduce the total quantity of water withdrawn from groundwater aquifers and to protect against contamination by reducing the rate at which contaminants can spread in the aquifer (e.g., excessive withdrawals from an aquifer located near the ocean can draw salt-water into the aquifer and contaminate wells).

7.8. Lining and Sealing Method of Groundwater Protection

Lining and Sealing

A German based Company, NaueFasertechnikGmbh& Co. (Anthonyet al., 1991) claimed that Carbofol ® (a registered brand product of them) offers complete lining systems for groundwater protection. They noted that NaueFasertechnik is the single source for complete groundwater protection. Single component sealing systems featuring Carbofolgeo membranes, are used in industrial facilities for storage, filling and transferring and for production, treatment and utilization - where there is a danger of groundwater contamination by hazardous materials.

When used in the construction of collection basins, Carbofol® high-density polyethylene (HDPE) geomembranes effectively protect groundwater and soil resources from pollution by potentially harmful contaminants. With a minimum thickness of 1.5 mm, Carbofol HDPE geomembranesfulfil a wide range of chemical resistance as well as physical durability requirements. In addition to their long-term chemical resistance, Carbofol® HDPE geomembranes have long-term UV resistance and are resistant to rodents as well as root growth when manufactured with a minimum thickness of 1.0 mm. By virtue of Carbofol's® high flexibility, more complicated connections are feasible without sacrificing quality or performance. Due to their advantageous melt index properties, Carbofol® geomembranes can be welded under a wider range of conditions, producing gas and watertight overlaps. Quality control tests are to be conducted continuously on every project in addition to the standard manufacturing quality control practices.

Protection

To ensure the performance and quality of a single component sealing system, mechanically bonded Secutex® or Depotex® protection nonwovens should be used above and below the Carbofol® geomembrane, the selection of which should be based upon project specific conditions. With a low gravel content sandy subsoil (< 30% gravel portion), a 400 g/m² mechanically bonded polypropylene (PP) Secutex® protection nonwoven is sufficient. However, with coarser grained soils, mechanically bonded Secutex® protection nonwovens with a mass per unit area of up to 1,200 g/m² should be considered. Alternatively, if the sealing system must perform in an aggressive chemical environment, Depotex® HDPE nonwovens should be used for the protection components to ensure long-term protection.

7.9. Some General Guidelines/Solution Options for Specific Problems

1. Over-Abstraction

Pumping fresh groundwater at too high a rate, or from too concentrated an area, usually from individual wells or sections of well-fields, causing localized salinity increases. The practical solution consists of careful management of well-fields and monitoring of the output and salinity of individual wells or trenches. Given enough time, the effects of over-abstraction can be reversed.

2. Physical Disturbance

Salt water invades areas which were previously fresh through the construction of marinas, canals, and waterways which are connected to the sea. The damage to the fresh groundwater is usually permanent. Also included in this category is any excavation below the water table, such as borrow pits for road construction, which expose open water and lead to salinization through evaporation. The practical solution includes impact assessment of plans for new marina or waterfront developments, approving designs, and inspecting construction. Also, avoiding excavation below the water table, if possible.

3. Point-Source Pollution

Specific incidents or local sources of pollution such as oil spills, leaks from underground storage tanks at gas stations, engineering workshops, chemical spills, etc. The practical solution is good working practices, approving designs for fuel and chemical storage and handling facilities, then inspecting construction and monitoring operation.

4. Solid Waste Disposal

Pollution from leachates from landfills, sludge disposal, and illegal dumping. This is really another form of point-source pollution, but has been separated out because it is usually regulated by different legislation. The practical solution is good working practices, ensuring landfills are correctly designed and constructed, and public education on illegal dumping.

5. Disposal Wells

Pollution from disposal or drainage wells which have been badly constructed, wrongly sited, or drilled to the wrong depth. Also inadequate treatment of waste before disposal down wells. The problem here is not with the concept of using disposal wells, which can be a safe, effective and low-impact method of disposal for liquid waste. The threat is from bad design, bad construction, or bad operation. The practical solution is therefore approving designs, inspecting construction, and monitoring operation.

6. Septic Tanks

Pollution from septic tanks, cesspits and latrines which have been badly constructed, built in the wrong place (such as below the water table), not emptied often enough, or not equipped with an accompanying disposal well. The practical solution is approving designs, inspecting construction, and public education.

7. Abstraction Wells

Abstraction wells (mostly private wells) which have been badly constructed, wrongly sited, or drilled to the wrong depth. Again, the practical solution is approving designs, inspecting construction, and public education.

8. Diffuse Pollution

Pollution over wide areas from poor use of fertilizers, chemicals and manures in agriculture, on golf courses, sports fields and public parks, poor use of treated effluent for irrigation. The practical solution is good working practices, control of products and public education.

The following guidelines can help to protect groundwater and water-well:

- Maintain your well and test the water quality annually.
- Keep household chemicals, paint and motor oil away from your well and dispose of them properly by taking them to a recycling center or household hazardous waste collection site.

- Limit your use of pesticides and fertilizers.
- Install a well cap and keep it clear of leaves, mulch, dirt, snow and other materials.
- Use caution when mowing around your well so you don't damage the well casing.
- Practice water conservation measures in your home and install low water use appliances.

7.10. Opportunities to Improve Groundwater

There are technological advances on the horizon that may offer opportunities to improve groundwater. These include:

- man-made wetlands
- terraced, overland flow systems
- package plants
- sand-filter systems, and
- greenhouse, peat, and bio-filter systems.

7.11. Challenges in Groundwater Protection and Management

Groundwater protection program faces substantial technical challenges. There is considerable uncertainty about the contamination sources themselves. The pathways for transport of contaminants from their sources may include four different hydrologic regimes: (1) surface streams and runoff, (2) near-surface groundwater, (3) intermediate-perched groundwater in the unsaturated (vadose) zone, and (4) a deep, regional aquifer. Each of these regimes adds considerable uncertainty to the understanding of the overall system. Even with best efforts to understand contaminant sources and pathways, the uncertainty will always be great.

Groundwater protection will become increasingly important as population densities continue to increase. In a watershed, contaminated groundwater has a potentially devastating effect. As a result, maintaining appropriate densities of development and proper disposal of sanitary sewer wastes are critical factors in ensuring the adequacy and quality of domestic water sources. Lowering of water-table can have a range of geo-chemical changes (e.g., arsenic contamination by oxidation process), which may pose a threat to groundwater.

Relevant Journals

Applied Geochemistry
Asian J. of Water Environmental Pollution
Applied Water Science
Air, Soil and Water Research
Australian J. Agril Engg

Australian Journal of Soil Research
British Journal of Applied Science & Technology
Engineering Geology
European J. of Agronomy Agriculture, ecosystem & environment
Geological Sci. and Technology Information
Geological Sciences Journal
Global J of Env Sci & Tech
Global advanced research journal of environmental science and toxicology
Groundwater
Groundwater Hydrology
Hydrological Processes
International Journal of Agricultural Sustainability
International Journal of Environmental Science and Toxicology (IJEST)
International Journal of Research And Innovations in Earth Science
International Journal of Water Resources Development

Questions

1. What are the elements of groundwater protection program?
2. Narrate the criteria commonly used for zoning the groundwater for protection.
3. Discuss the measures for controlling quality degradation.
4. Mention the elements of Wellhead protection programs
5. Discuss the need and issues of laws and regulations in water quality protection
6. Briefly narrate the regulatory instruments in controlling water quality
7. Do you think that Community involvement is an effective way of water quality protection? Discuss.
8. Discuss the Challenges and opportunities in groundwater protection and management

References

Anthony, W. E., Boschuk, J. & Koerner, R. M.(1991). Prefabricated bentonite clay liners. *Geotextiles and Geomembranes*, *10*(5-6), 575-599.

Chave, P., Howard, G., Schijiven, J., Appleyard, S., Fladerer, F. & Schimon, W. (2006). Groundwater protection zones. IWA Publishing, London, UK.

Forster, C. F., Fernandez, N. & Chacin, E. (2000). The use of ultrasound to enhance the thermophilic digestion of waste activated sludge. *Environ. Technol.*,*21*, 357–362.

Chapter 8

Economics of Water Resources Management

Abstract

An investment or intervention is said to be economically efficient when it maximizes the value of output from the resources available. Economic analysis of water use includes the values generated by production activities, the opportunity costs of inputs, and any pertinent costs or benefits that are external to producers and consumers. Economic analysis is helpful in identifying opportunities to increase the net values generated with limited water resources by various techniques/alternatives, and in designing policies that encourage farmers and water agency personnel to improve water management practices in ways that enhance social net benefits. In selecting the best water resources project (water supply/irrigation/drainage/flood control) among alternatives, or to judge a single project for its economic viability, several criteria of economic indices are used, which have been described in this chapter with a detail analytical procedure and sample examples.

8.1. Economic Aspects in Water Resources Projects

Irrigated areas constitute about 17% of the total farmland in the world and provide 50% of the global agricultural produce. In the Middle Eastern countries, irrigated fields yield 70% of the agricultural produce. An irrigation system as an object of the economy depends on the economic situation, and the market for agricultural commodities, and, at the same time, exerts its own impact on these factors. The economic state of an IS is determined by its budgetary balance. The main economic target of ISs is to render paid services to agricultural water consumers (AWC) by delivering high-quality fresh water at the prescribed time and place, and in the required volume. The efficiency of these services is evaluated for working, redesigned, and newly constructed ISs on the basis of the economic design, with allowance for current-outlay and capital costs, as well as for proceeds from services rendered. Under conditions of paid water consumption, the role of operational gauging services increases, as well as the need for more accurate water assessment, because any supply shortfall is fraught with disbenefit. ISs must be ready both to supply their clients with water in the event of failure of traditional water sources and to compensate AWCs and other water consumers for their losses. For example, if enhanced extraction of groundwater results in decrease of the water table and deterioration of the environment, the IS should perform the remedial work. The IS must also pay water consumers for any shortfall caused by disturbance of the water regime of source water reservoirs.

8.2. Economics of Irrigation Systems

Irrigated farming is a traditional agricultural practice developed over the centuries in countries with a hot climate. Many generations have expended important material and financial resources for the development of irrigation. In droughty regions, artificial increase of soil moisture content (using irrigation systems) is frequently the only measure capable of meeting the demands of the local population and industry for crop and animal products. Land irrigation improves the water-salt balance of the soil and sustains its economic fertility. Irrigation farming involves a higher level of agricultural production standards, as compared to dry farming. Its specificity is the use of two principal means of production- land and water, in a single technology of crop production. From the viewpoint of economics, the increase in productivity in irrigation farming is due to capital investment into the fertility of irrigated areas.

An irrigation system (IS) is a technological synthesis of water management and agriculture, realized through transportation of water from a source to an irrigated field. Therefore, the economics of IS is closely related to the economics of water management and agriculture.

Mutually acceptable (negotiated) price of water is found by seeking mutual compromises or other sources for reimbursement of IS and AWC costs. A consensus may be reached at the expense of profit. A unified price of water (from a water source) is generally fixed for the whole water-management basin. In a project, the study and analysis of commercial profitability for each crop are performed for both current and design conditions with allowance for the run of the market, inflation rate, development of equipment and technologies, and increasing material inputs, etc.

8.3. Water as an Economic Good

Water has traditionally been regarded as a public good to which no one can be denied access, except some dry areas of the world. Water was available from rainfall, streams, wells, tubewells (bored into aquifers), and diversion of surface flows. Water provision was the responsibility of local people and the investments for this purpose were small. But the situation has been changed. There has been a gradual realization that global stocks are limited and that supply augmentation cannot indefinitely meet ever-increasing demands; and supplies are degraded (polluted and depleted). The growing scarcity and rising cost of water have led to the realization that water has to be allocated and used more efficiently, and to be treated as an economic good. Markets and prices can be used to help ensure sustainable usage, minimize wastage, ensure efficient allocation, and provide incentives for the development of water-efficient technologies, re-use and recycling.

There may be significant opportunities to increase the net values generated with limited water resources by various techniques/alternatives (improving the distribution of water among farmers, reducing the negative/off-farm effects of irrigation and drainage activities, adopting deficit irrigation, growing low water demand crops, high value (price) crop cultivation, amongst others). Economic analysis is helpful in identifying those opportunities and in designing policies that encourage farmers and water agency personnel to improve water management practices in ways that enhance social net benefits.

8.4. Application of Economic Principles in WRs Fields

In economic and financial decision making, we need the following types of project decision for which criteria are needed:

(i) Testing the economic viability of a project
(ii) Choosing the best option among project alternatives, and
(iii) Choosing the least-cost option for achieving the targeted (or same) benefit

8.5. Significance of Economic Analysis in WRs Projects

Where resources are scarce, proper planning and decision-making at different levels is essential. In land-limiting situation, maximizing yield per unit of land, and not yield per unit of water, is a more viable objective. In water-scarce areas, water, not land, is the primary limiting factor to improve agricultural production. Accordingly, maximizing yield per unit of water, and not yield per unit of land, is a more viable objective for on-farm water management. It is useful to know how to best utilize the insights and instruments of economics in the design and management of irrigation & other development schemes and interventions.

There may be significant opportunities to increase the net values generated with limited water resources by various techniques/alternatives (improving the distribution of water among farmers, reducing the negative/off-farm effects of irrigation and drainage activities, adopting deficit irrigation, growing low water demand crops, high value (price) crop cultivation, amongst others). Economic analysis is helpful in identifying those opportunities and in designing policies that encourage farmers and water agency personnel to improve water management practices in ways that enhance social net benefits.

In many practical problems, we, the engineers, faces multiples alternatives in selecting a development plan or project. In such situation, we have to decide the right thing (the best one) – the technology, the structure, project adoption and implementation, etc. among alternatives. Economic merit of a project is judged by comparing projected benefits with projected costs. Under normal condition, an irrigation or drainage development project should be technically feasible, economically viable, and environmentally sustainable. So, we have to consider technical performance, durability, financial involvement (input), outcome/output, environmental aspects, etc. Economic analysis of income and services producing from investment is thus essential. In engineering activities, or activities having a major role of engineers, the engineers are best fitted to analyze such thing due to their background in engineering principles.

The role of economic analysis in decision making process is well established. Economic analysis of water use includes the values generated by production activities, the opportunity costs of inputs, and any pertinent costs or benefits that are external to producers and consumers. Economic analysis can be helpful in describing the private (internal) and public (external) costs of an inefficient allocation of resources, and in determining strategies for moving towards an efficient allocation.

As the resources are becoming scarce and have alternate use among different sectors (agriculture, industry, fisheries, urban use), evaluation of different alternates is pre-requisite to achieve economic efficiency. Economic efficiency is achieved when limited resources are used in a manner that generates the greatest net value.

8.6. Basics of Economic Decision Tools

Cost and Revenue

Elements of Cost and Revenue

Total cost of a development project (or production of a particular item) can be divided into two portions:

- Fixed cost, and
- Variable cost

Fixed cost is the portion of the total cost that requires regardless of the amount of production (small or greater). Examples are rent of office building, salary of security personnel, etc. Variable cost is the rest portion of the total cost that varies with the amount of production. It increases with the increase in production, and decreases with the decrease in production. Examples are labour cost, interest on running capital, etc. For economic production process, one must use the least costly combination of inputs for a particular level of output. The optimum mix of output achieves for a given level of benefits at least cost, or in other word, the maximum level of benefit for a given level of cost.

In economic terms, the cost of water has two broad components:

- the cost of its provision (including both fixed (investment) and variable costs (operation and maintenance), and
- its opportunity cost, or the production value forfeited/offered in alternate use

These costs are real and unavoidable, and someone has to pay - the user, the taxpayer or future generations.

The revenue from a project (or production system) is the total earnings *or* gain expressed in monetary form. The net benefit is the difference between total cost and total income *or* revenue (i.e., net benefit = total revenue - total cost). The vertical difference between cost curve and revenue curve represents the net financial benefit. Maximum net benefit can be determined by constructing cost and revenue curve, and measuring the vertical difference between them. In a free market economy, productively efficient firms use these curves to find the optimal point of production, where they make the profit most.

Factors Affecting Cost and Return

All the cost components can affect the cost. In addition, the following factors can affect the cost and return:

- Time
- Inflation/Deflation

8.7. Cost Estimation

Forecasting the Benefits and Costs

Accurate forecasting of future costs and benefits can be the most difficult and critical step in project financial analysis. All possible costs and benefits of the proposed project should be taken into account, including non-monetary costs and benefits.

The costs should include:

- Input cost
- Operational cost
- Maintenance cost
- Opportunity cost of the inputs/resources
- Uncertain cost

Input Cost

All sorts of input costs (both fixed and variable) should be considered in analysis. Salaries of the farm manager and other staffs, fee of the consultant, etc. fall under this category.

Opportunity Cost

The opportunity cost of a project is the potential benefits that are lost by selecting the project. More specifically, it is equal to the net benefit that is to be achieved from alternate use of the resources (capital, human resources, or other major inputs). For example, a project has been proposed to develop an agricultural farm for crop production over a land of 500 hectares. If the farm is not developed, the farmer will get 8000 US$ per annum as hire for cattle grazing without any investment. Now, $8000 is to be added to the total cost as opportunity cost. If this opportunity cost is not included as cost in project analysis, then some proposals may appear to be better because they are using existing resources free.

Operational Cost

This includes salary of the laborers, fuel for the tractor, fertilizers, insecticides, etc.

Maintenance Cost

This category includes repair cost of the farm instruments, computers, cars, etc.

Uncertain Cost

Sometimes it is difficult to estimate the cost because they are dependent on an unpredictable environment. In such uncertain case, still it is possible to make an estimate (expected value).

8.8. COST ANALYSIS – ECONOMIC DECISION TOOL

Decision-makers are often faced with the challenges of resource allocation. Resources are scarce; therefore, they must be allocated judiciously. Resource-allocation decisions should be guided by considerations of cost in relation to expected benefits.

Different types of cost analysis are used in program *or* project selection. Cost-effectiveness analysis, and cost-benefit analysis are commonly used in program evaluation and selection.

Cost Effectiveness Analysis

In some cases, the benefits or outcomes of a projects can not be valued (or not directly measurable), but can be quantified or have a specific outcome. For example, in a pollution control project the outcome is pollution control (having specific quality or grade), with a specific cost.

A measure of the cost effectiveness is obtained by measuring costs against outcome (average incremental economic cost). The cost effectiveness ratio – the cost per unit change in quality (improving certain quality, lessening hazardous effect, etc.) for each of the alternative methods are compared. The technique or method that needs minimum cost to attain targeted outcome is most cost effective. Analysis of cost effectiveness allows choosing the best one evaluating various options.

There may be circumstances where the project alternatives have more than one outcome. In order to assess the cost effectiveness in such a case, it is necessary to devise a testing system where the outcomes of different dimensions can be added together. It is also necessary to select some weights for adding the different dimensions reflecting their relative importance in relation to the objectives of the project. Such a use of cost-effectiveness analysis is called weighted cost-effectiveness analysis.

8.9. Financial Analysis of Water Resources Projects

8.9.1. Relevant Terminologies

Benefit-Cost Ratio (BCR) for Multi-Year Project

For multi-phase or multi-year project, bcr is defined as the ratio of the present value of the economic benefits stream to the present value of the economic costs stream, each discounted at the economic opportunity cost of capital the ratio should be greater than 1.0 for a project to be acceptable.

Present and Future Value

Present value tells the current worth of a future sum of money. *Future value* gives one the future value (at a future date) of cash that one have now. For a multi-year project, future value of the present values (and near future cash flows, if any) are calculated at the end of the project life.

Initial Investment

It is the investment made at the beginning of a project. The initial investment may include hardware (cost of buying and installing the machine, structural setup), software, startup costs, licensing fees, etc. Since most of the projects involve initial cash outflow, its value is usually taken as negative.

Salvage Value

It is the money that can be gained by selling the project inputs (hardware or instruments, properties, etc.) at the end of the project. If there is, the amount would be added to the income amount, normally at the last year of the project.

Nominal and Real Value (for Cost, Benefit, Interest Rate)

The real value is one where the effects of inflation have been factored/considered; whereas, a nominal value is one where the effects of inflation have not been accounted for. To account for inflation, either real or nominal values may be used, but with consistently. That is, nominal costs and benefits require nominal discount rates, and real costs and benefits require real discount rates.

The real interest is one where the effects of inflation have been factored in. a nominal interest is one where the effects of inflation have not been accounted for. That is,

$$\text{Real interest rate} = \text{Nominal interest rate} - \text{Inflation}$$

Inflation is generally positive. If so, the real interest rate is lower than the nominal interest rate. If the inflation rate is negative (i.e., deflation), then the real interest rate will be lager than the nominal interest rate.

Income Stream or Cash Flow

An income stream is a series of amounts of money. Each amount of money comes in or goes out at some specific time, either now or in the future. Sample example of an income stream is given below:

Year	0	1	2	3	4	5
Income amounts	-$5000	$500	$600	$2000	$3000	$600

The net cash flow for each year of the project is the difference between revenue (or benefit) and cost. All the amounts in the income stream are net income, meaning that each is revenue minus cost, or income minus outgo.

For simplicity, the cash flow in the above example has at one-year intervals, but real time investments can have cash flow (income minus expenses) at irregular times. But the principles of evaluation are the same for all cases.

Internal Rate of Return (IRR)

Internal rate of return is traditionally defined as the discount rate at which NPV is equal to zero. It is the projected discount rate that makes the NPV of a project equal to zero. It is one of the capital budgeting methods used by firms to decide whether they should make long-term investments. It helps making decision about which project is most economical for a particular situation.

Discounting

Discounting is a procedure developed by economists to evaluate investments that produce future income. It is the process by whereby the values of future effects are adjusted to render them comparable to the values place on current costs and benefits. It is accomplished by multiplying the future value(s) by discount factor.

Discount Rate

A percentage rate representing the rate at which the value of equivalent benefits and costs decrease in the future compared to the present. The discount rate is used to determine the present value of the future benefit and cost streams.

The rate can be based on the alternative economic return in other uses given up by committing resources to a particular project, or on the preference for consumption benefit today rather than later. The economic net present value is generally calculated for each project alternative using the Bank interest rate.

Compounding

It is the process whereby the values of present effects are adjusted to render them comparable to the values placed on future consumption, costs, and benefits.

Discounted Cash Flow

The discounted cash flow approach describes a method to value a project or an entire company using the concepts of the time value of money. It is widely used in investment finance, real estate development, corporate financial management and other development projects. All future cash flows are estimated and discounted to give them a present value. Discounted cash flow (DCF) includes the present value (PV) and the internal rate of return (IRR) methods of analyzing cash flows. The DCF provides insight into financial management.

The discounted cash flow formula is derived from the future value formula for calculating the time value of money and compounding returns.

The discounted cash flow formula is derived from the future value formula for calculating the time value of money and compounding returns.

$$FV = PV.\ (1+i)^{n} \tag{8.1}$$

Where,

FV = future value

PV = present value

i = interest rate (or time value of money)

n = number of year

For single cash flow in one future period, discounted cash flow equation can be expressed as:

$$DPV = \frac{FV}{(1+r)^n}$$

where,

DPV = discounted present value of the future cash flow (FV)

r = discount rate

In case of cash flows in multiple time periods, discounted present value is the sum of the individual present values, expressed as:

$$DPV = \sum_{t=0}^{N} \frac{FV_t}{(1+r)^t} \tag{8.2}$$

Where, t = time period (year number from the present year)

8.9.2. Expression of Cost and Benefit

In different projects, investment (or input requirement) and the return (or outcome) do not follow same time. To overcome this problem, the inputs and outputs are converted to an equivalent amount.

There are different ways to express and compare costs and benefits that occur in multiple time periods on a consistent basis. Procedures for calculating cost and benefit that occur in more than one time period are:

(a) Net present value (NPV)
(b) Net future value (NFV)
(c) Annualized value

For NPV, discounting places all costs and benefits at the present time period; annualization spreads them smoothly over time; and for NFV, accumulation places them at a future time. Among the three procedures, net present value method is widely used.

Net Future Value

Net future value (NFV) is a way of rendering cost and benefit that occur in more than one time period. Net future value is the total future value of all cash flows. It is an estimate of what the principal will become over time.

The NFV value of a projected stream of current and future costs and benefits is obtained by multiplying the benefits and costs in each year by a time dependent weight, d_t, and adding all of the weighted as follows:

$$NFV = d_0NB_0 + d_1NB_1 + d_2NB_2 + ... + d_{n-1}NB_{n-1} + d_nNB_n \quad (8.3)$$

Where, NB_t is the difference between benefits and costs that accrue in year t, and the accumulation weights (also termed as compound factor), d_t, is given by:

$$d_t = (1+r)^{(n-t)} \quad (8.4)$$

where, r is the discount rate. Thus, the equn. (11.3) can be written as:

$$NFV = \sum_{t=0}^{n} NB_t (1+r)^{(n-t)} \quad (8.5)$$

Annualized Values

For example, in the absence of discount rate, a cost of US$ 1,000 becomes US$ 2,000 at the end of first year, US$ 3,000 at the end of second year, US$ 4,000 at the end of third year. From that, it can be said to cost US$1,000 per year in annualized costs over the three years period.

Costs and benefits may be annualized separately by using a two step process. Comparing annualized costs to annualized benefits is equivalent to comparing the present values of costs and benefits. To compare average cost effectiveness among alternative policy options, divide the annualized cost by annualized benefit.

Net Present Value

Net present valueis the difference between the present value of the benefit stream and the present value of the cost stream for a project. The word '*net*' in '*net present value*' indicates that the calculation includes the initial costs as well as the subsequent cost and benefit.

Costs and benefits of a development projects or a policy frequently occur at different times. The NPV of the time-phased costs over the economic life of an investment project is the best single-number measure of its life-cycle cost.

If we assume that income comes (benefit) or goes (cost) in annual bursts, the net present value (NPV) value of a projected stream of current and future costs and benefits is obtained by multiplying the benefits and costs in each year by a time dependent weight, d_t, and adding all of the weighted as follows:

$$NPV = NB_0 + d_1NB_1 + d_2NB_2 + \ldots\ldots\ldots\ldots + d_{n-1}NB_{n-1} + d_nNB_n \quad (8.6)$$

$$= NB_0 + \sum_{t=1}^{n} d_t NB_t$$

Where, NB_t is the difference between benefits and costs that accrue in year t, n is the final period in the future (n= t), and the discounting weight, (also termed as discount factor), d_t, is given by:

$$d_t = 1/(1+r)^t$$

where, r is the discount rate. Thus, the equation can be written as:

$$NPV = NB_0 + \sum_{t=1}^{n} \frac{NB_t}{(1+r)^t} \quad (8.7)$$

As the initial investment is a cost, it is treated as negative benefit.
That is, NPV = (- Initial investment) + Σ (Net cash flows at year t)/$(1 + r)^t$

8.9.3. Analysis of Net Present Value (NPV)

Analysis of net present value requires four basic steps:

- forecast the benefits and costs in each year.
- determine a discount rate.
- use a formula to calculate the NPV

Calculation of Net Present Value

At first, calculate the net benefit for each year by subtracting the cost from the benefit. Then, calculate the net present value for the net benefit of each year using the formula given below:

$$NPV_n = \frac{NB_n}{(1+r)^n} \quad (8.8)$$

where,

NPV_n = net present value for the year n
NB_n = net benefit for the year n

n = chronological serial number of the year in the cash flow series for which the NPV is calculating

In NPV calculation, for simplicity, it is assumed that the discount rate will not change over the life of the project.

Sum up the NPVs to get total NPV.

The main limitation of NPV analysis is the difficulty of accurately forecasting future costs and benefits. A positive net present value means the investment is better. A negative net present value means the alternative investment, or not borrowing, is better.

$$F = P(1 + r)^N \tag{8.9}$$

That is, we can say,

Future value of any present amount = (Present amount) × (1 + rate of return)^N

Where, N is the number of years into the future that the income amount will be received.

Assume an income stream where m is the number of payment per year. In this case, annual rate of return (r) can be split as r/m, and the present value can be computed as:

$$F = P(1 + r/m)^{Nm} \tag{8.10}$$

8.10. Indicators and Decision Rules for Project Selection

8.10. 1. Indicators for Project Judgment

A test of economic viability needs to be applied for single project as well as multiple options. The following indicators are normally used to compare the project alternatives:

(i) Benefit-cost ratio (BCR)
(ii) Net present value (NPV)
(iii) Economic internal rate of return (IRR)

Comparative Merits and Demerits of the Indicators

The BCR is a direct indicator of benefit (gain) or loss from a project. However, it does not tell us about the total income from the project alternatives; thus not suitable to compare different sized projects and mutually exclusive projects. At the same BCR (or even at lower BCR), the total income from a large project may be much higher than that of a small project.

The NPV is suitable for absolute measure of value of an inflow-outflow system. The NPV is well accepted for sound reasons, but it has limitations. To solve for NPV, one must first calculate the opportunity cost of the capital, also called 'discount rate.' The rate is used to calculate NPV. The NPV is highly sensitive to the discounting rate. Using different discounting rates can change NPV ranking, and therefore not very useful for comparisons between organizations – especially those of different sizes projects.

The IRR is a rate or ratio, not an amount, thus more useful for comparing unlike/dissimilar investments. It is also useful for making comparisons between different sized farms and between different periods.

Both IRR and NPV are widely used to decide which investments to undertake and which investments not to make. The aim is to maximize the total NPV, but not to IRR. The major difference between IRR & NPV is that while NPV is expressed in monetary units (Dollar, for example), the IRR is the true interest rate expected from an investment expressed as a percentage.

In conclusion, it can be said that the BCR and IRR give identical signal, and hence have essentially equivalent utility. The two measures complement each other. The NPV is a better absolute measure, where IRR & BCR are better relative measure. When a single indicator is used, the NPV and IRR can rank the alternatives differently. The BCR, IRR and NPV together can give a better picture of the problem than either alone.

8.10.2. Determination of the Indicators

The BCR and NPV have been discussed earlier.

IRR

The internal rate of return (IRR) is the discount rate for which the NPV of the project cash flow is zero. It is determined by trial, determining NPV at different discount rate, as of Figure given below. It can easily be done in Microsoft Excel (Figure 8.1).

The curve crosses the horizontal line (indicating NPV = 0) between 0.04 and 0.06. So the internal rate of return is between 0.04 and 0.06 (nearly at 0.05). Thus, the internal rate of return (IRR) is 0.05.

8.10.3. Selection Criteria and Decision Rules

8.10.3.1. Single Project Evaluation

For single project, the following economic criteria should be fulfilled:

- BCR should be greater than one (i.e., BCR > 1)
  Or,
- The NPV should positive (i.e., NPV > 0)

Decision Procedure

Accept/implement the project if BCR > 1, or NPV > 0.

Other aspects such environmental impacts, basic need of the people, employment opportunity, proper use of local inputs/materials, political will, etc. may also be considered in decision making.

8.10.3.2. Multiple Project Evaluation

(A) Mutually Independent Projects

'Mutually Independent projects' are those projects which (any one) can be selected without any restriction on the others.

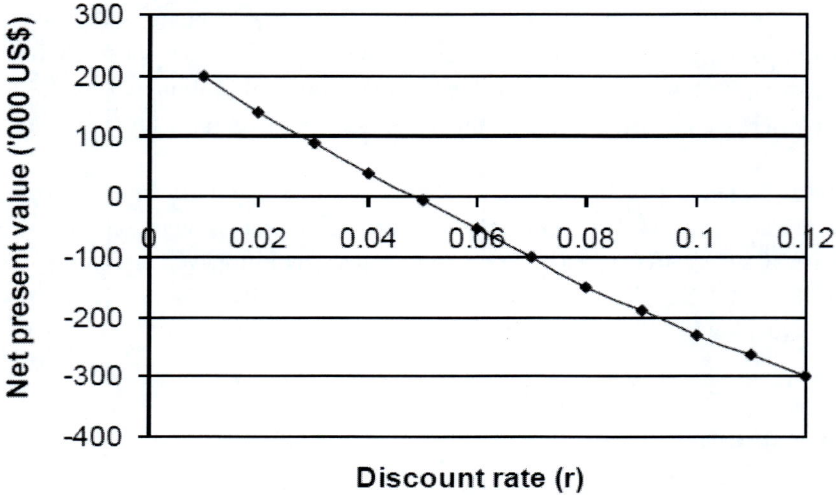

Figure 8.1. Process for determining IRR.

Essential Selection Criteria

- the NPV should be positive
- the BCR should be greater than unity

Selection Procedure

Choose the project first having highest BCR and IRR. Then, select second project which have second highest BCR and IRR. If fund is available, select sequentially all projects having BCR > 1 and NPV > 0.

(B) Mutually Exclusive Projects

'Mutually exclusive projects' are those projects from which any one can be selected with sacrificing of the others (that is, only one can be selected).

Essential Criteria

- the BCR should be greater than unity
- the NPV should be positive

Selection Procedure

Choose the project having highest NPV.

Other Considerations

In addition to economic criteria, other factors such as governmental/public priority future well-being of the citizen, self-sufficiency of the basic needs, environmental factor, etc. should be taken into account. Even a project is not financially profitable (for example - basic need of the people, such as drinking water supply), it should be implemented for the interest of the people.

8.11. Workout Problems

Example 8.1

A flood mitigation reservoir has been proposed near Calkata, India. The initial cost of construction is estimated as 9×10^4 US\$. The operating and maintenance cost is about 3,000 US\$ per year. Expected benefit from 'prevention of flood damage' is about 1.8×10^4 US\$ per year. Assume that no major dragging will be necessary within 15 years (that is effective life is 15 years). Is the project economically justifiable? Assume a discount rate of 8%.

Solution

Discount rate, r = 8%

Total cost for a year = construction cost + maintenance & operational cost

Net benefit = benefit – cost

Present value of "net benefit" for a particular year, $PV = F/(1 + r)^N$

The calculations are summarized below:

Yr	Construction cost (US$)	Maintenance & Operation cost (US$)	Total cost (US$)	Benefit (US$)	Net benefit (US$)	PV of net benefit for yr *i* (US$)
0	9×10^4	0	9×10^4	0	-90000	-90000
1		3000	3000	1.8×10^4	15000	13888.9
2		3000	3000	1.8×10^4	15000	12860.1
3		3000	3000	1.8×10^4	15000	11907.5
4		3000	3000	1.8×10^4	15000	11025.4
5		3000	3000	1.8×10^4	15000	10208.7
6		3000	3000	1.8×10^4	15000	9452.5
7		3000	3000	1.8×10^4	15000	8752.4
8		3000	3000	1.8×10^4	15000	8104.0
9		3000	3000	1.8×10^4	15000	7503.7
10		3000	3000	1.8×10^4	15000	6947.9
11		3000	3000	1.8×10^4	15000	6433.2
12		3000	3000	1.8×10^4	15000	5956.7
13		3000	3000	1.8×10^4	15000	5515.5
14		3000	3000	1.8×10^4	15000	5106.9
15		3000	3000	1.8×10^4	15000	4728.6
			Sum			38392.18

NPV of benefit = 38392.18

Since the NPV is positive, the project is justified economically. (**Ans.**)

Example 8.2

In 'North China Plain Area' of China, the estimated cost of a Drainage Project is US\$ 870,000 (at the beginning of the project). The effective life of the project is expected to 30 years. The interest on capital is 6%. The annualized value of the operation and maintenance cost is 2% of the initial cost. The salvage value from the project is US\$5000 at the end of the project. The annualized value of the benefit from the project is US$ 56,430. Is the project economically viable?

Solution

Given:
Initial cost = 870,000 US$
Effective life = 30yrs
Interest rate = 6%
Annual. Opera. Cost = 2% of initial cost
Salvage value at yr 30 = 0 US$
Annual benefit = 56,000 US$

$$\text{We know, annualized value, } A = \frac{F \times r}{(1+r)^N - 1} \qquad \text{(i)}$$

Putting the values, Annualized value of initial cost = 63205 US$
Annualized ope. & mainte. cost = (2/100) × 63205 = 17400 US$
Total annual cost = 63205 US$ + 17400 US$ = 80605 US$
Total annual benefit = 56,000 US$

BCR = (Total annual benefit)/(Total annual cost)
= 56,000 US$/80605 US$
= 0.695

As the BCR < 1.0, the project is not justified. (**Ans.**)

Example 8.3

An irrigation development project has been planned in Cameron. The tentative cash-flow from the project is given below. Perform financial analysis and comment on the economic viability of the project. Assume discount rate of 6%.

Cash-flow

Year	Initial cost ($)	Operation cost ($)	Maintenance cost ($)	Revenue ($)
0	60,000	7000		12000
1		7000		12000
2		7000		12000
3		7000		12000
4		7000	4000	12000
5		7000		12000
6		7000		12000
7		7000		12000
8		7000		12000
9		7000		12000

Year	Initial cost ($)	Operation cost ($)	Maintenance cost ($)	Revenue ($)
10		8000	6000	12000
11		8000		11000
12		8000		11000
13		8000		11000
14		8000		11000
15		8000	5000	11000
16		8000		11000
17		9000		10000
18		9000		10000
19		9000		10000
20		9000		10000

Solution

Total cost for a year = Operation cost + Maintenance cost

(For the starting year (zero year), only the initial cost)

Net revenue = total revenue – total cost

Present value of "net revenue" for year N, $P = F/(1 + r)^N$

For sample illustration, present value (PV) of 'net revenue' for year 5,

$P_5 = 5000/(1+ 0.06)^5 = 3736.3$

The results are summarized below.

Year	Total cost ($)	Net revenue ($)	PV of net revenue ($)
0	67,000	-55000	-40500
1	7000	5000	4717.0
2	7000	5000	4450.0
3	7000	5000	4198.1
4	11000	1000	792.1
5	7000	5000	3736.3
6	7000	5000	3524.8
7	7000	5000	3325.3
8	7000	5000	3137.1
9	7000	5000	2959.5
10	14000	-2000	-1116.8
11	8000	3000	1580.4
12	8000	3000	1490.9
13	8000	3000	1406.5
14	8000	3000	1326.9
15	13000	-2000	-834.5
16	8000	3000	1180.9
17	9000	1000	371.4
18	9000	1000	350.3
19	9000	1000	330.5
20	9000	1000	311.8
		NPV =	**-17762**

The present value of the 'net revenue' is less than zero (-ve), thus the project is not economically justified. (**Ans.**)

Relevant Journals

Agricultural Economics
Agricultural Economics (AGRICECON)
Agricultural and Resource Economics Review
Agricultural Engineering
Australian J. of Agril. & Resource Economics
Australian Journal of Agricultural Economics
European Review of Agricultural Economics
Journal of Agricultural Applied Economics
Review of Agricultural Economics

EXERCISE

8.1 What are the indicators commonly used for project judgment/selection?
8.2 Discuss merits and demerits of different indicators.
8.3 Describe the procedure for determining different indicators.
8.4 Discuss the considerations for "Governmental (public)" and "Private" investment/ project.
8.5 Determine the economic rate of return for an investment of "Sinhoa Groups of Industries," from the following cash-flow of the project:

Year	0	1	3	5	7	9
Cost	50000	5000	0	1000	1500	1000
Revenue	0	10000	10000	40000	10000	20000

8.6 A project is planned by "Jamuna Group of Industries" to implement, and the tentative cash-flow of the proposed project is given below. The management of the company wants at least 12% return per year from the project. From the financial analysis, comment on the justification of the project.

Cash-flow of the project:

Year	0	1	3	6
Cash outflow ($)	50,000	0	0	0
Cash inflow ($)	0	10000	20000	80000

8.7 An investor is investing $20,000 today. If he receives $6000, 15000, and 16000 after 2, 3, & 4 year respectively, what is the rate of return?
8.8 Perform economic and financial analysis of the given projects to justify the investment and suggest the best project option, when they are:
(a) Mutually independent,
(b) Mutually exclusive.

Assume discount rate of 6%.

Project-A

Year	0	1	2	3	4	5
Cost ($)	70000	5000	15000	5000	10000	5000
Revenue ($)	0	12000	35000	25000	20000	25000

Salvage value at year 5 is $ 6000.

Project-B

Year	0	1	3	5	7
Cost ($)	55000	8000	1000	2000	2000
Revenue ($)	0	27000	45000	28000	22000

No salvage value.

Project-C

Year	0	1	2	3	5
Cost ($)	120000	5000	5000	5000	4000
Revenue ($)	0	68000	70000	60000	40000

No salvage value.

Chapter 9

STATISTICAL METHODS IN WATER RESOURCES

ABSTRACT

Higher or lower rainfall, or changes in its spatial and seasonal distribution would influence the spatial and temporal distribution of runoff, soil moisture and groundwater reserves, and would affect the frequency of droughts and floods. Further, temporal change in precipitation distribution will affect cropping patterns and productivity. Proper identification of flood seasonality is important for many practical applications in hydrology and water resources management.

9.1. GRAPHICAL ANALYSIS OF SINGLE DATA SETS

Graphs provide visual summaries of data which more quickly and completely describe essential information than do tables of numbers. Graphs are essential for two purposes:

1. to provide insight for the analyst into the data under scrutiny, and
2. to illustrate important concepts when presenting the results to others.

The first of these tasks has been called exploratory data analysis (EDA). EDA procedures often are (or should be) the 'first look' at data. Patterns and theories of how the system behaves are developed by observing the data through graphs. These are inductive procedures - the data are summarized rather than tested. Their results provide guidance for the selection of appropriate deductive hypothesis testing procedures.

Graphical methods summarizes the information for the analyst will also provide insight into the data for the reader or audience. Some of the graphical methods are described below.

9.1.1. Histograms

Histograms are familiar graphics, and their construction is detailed in numerous introductory texts on statistics. Bars are drawn whose height is the number ni, or fraction ni/n, of data falling into one of several categories or intervals. Iman and Conover (1983) suggested that, for a sample size of n, the number of intervals k should be the smallest integer such that $2k \geq n$. Histograms have one primary deficiency - their visual impression depends on the number of categories selected for the plot.

Histograms are quite useful for depicting large differences in shape or symmetry, such as whether a data set appears symmetric or skewed. They cannot be used for more precise judgements such as depicting individual values. However, histograms are excellent when displaying data which have natural categories or groupings. Examples of such data would include the number of individual organisms found at a stream site grouped by species type, or the number of water-supply wells exceeding some critical yield grouped by geologic unit.

9.1.2. Quantile Plots

Quantile plots visually portray the quantiles, or percentiles (which equal the quantiles times 100) of the distribution of sample data. Quantiles of importance such as the median are easily discerned (quantile, or cumulative frequency = 0.5). With experience, the spread and skewness of the data, as well as any bimodal character, can be examined. Quantile plots have three advantages:

1. Arbitrary categories are not required, as with histograms or S-L's.
2. All of the data are displayed, unlike a boxplot.
3. Every point has a distinct position, without overlap.

9.1.2.1. Construction of a Quantile Plot

To construct a quantile plot, the data are ranked from smallest to largest. The smallest data value is assigned a rank I = 1, while the largest receives a rank i=n, where n is the sample size of the data set. The data values themselves are plotted along one axis, usually the horizontal axis. On the other axis is the "plotting position," which is a function of the rank i and sample size n.

Quantile plots are sample approximations of the cumulative distribution function (cdf) of a continuous random variable.

Plotting Positions

Variations of quantile plots are used frequently for three purposes:

1. to compare two or more data distributions (a Q-Q plot),
2. to compare data to a normal distribution (a probability plot), and
3. to calculate frequencies of exceedance (a flow-duration curve).

Different plotting positions have traditionally been used for each of the above three purposes. It would be desirable instead to use one formula that is suitable for all three. Numerous plotting position formulas have been suggested, most having the general formula:

$$p = (i - a)/(n + 1 - 2a)$$

where a varies from 0 to 0.5.

Five of the most commonly-used formulas are:

Reference	Formula
Weibull (1939)	i/(n + 1)
Blom (1958)	0.375 (i −0.375)/(n + 0.25)
Cunnane (1978)	0.4 (i −0.4)/(n + 0.2)
Gringorten (1963)	0.44 (i −0.44)/(n + 0.12)
Hazen (1914)	0.5 (i −0.5)/n

The Weibull formula has long been used by hydrologists in the United States for plotting flow-duration and flood-frequency curves. It is used as the standard reference for determining flood frequencies in the United States (Interagency Advisory Committee on Water Data, 1982). The Blom formula is best for comparing data quantiles to those of a normal distribution in probability plots, though all of the above formulas except the Weibull are acceptable for that purpose (Looney and Gulledge, 1985). The Hazen formula is used by Chambers and others (1983) for comparing two or more data sets using Q-Q plots. Separate formulae could be used for the situations in which each is optimal.

Cunnane (1978) presents the arguments for use of this formula over the Weibull when calculating exceedance probabilities.

9.1.3. Probability Plots

Probability plots are used to determine how well a data set fit a theoretical distribution, such as the normal, lognormal, or gamma distributions. This could be attempted by visually comparing histograms of sample data to density curves of the theoretical distributions. However, research into human perception has shown that departures from straight lines are discerned more easily than departures from curvilinear patterns. By expressing the theoretical distribution as a straight line, departures from the distribution are more easily perceived. This is what occurs with a probability plot.

To construct a probability plot, quantiles of sample data are plotted against quantiles of the standardized theoretical distribution. For a given cumulative frequency (plotting position, p), quantiles from each curve are paired and plotted as one point on the probability plot. Note that quantiles of the data are simply the observation values themselves, the pth quantiles where $p = (i - 0.4)/(n + 0.2)$. Quantiles of the standard normal distribution are available in table form in most textbooks on statistics.

Specialized ‘probability paper’ is often used for probability plots. This paper simply retransforms the linear scale for quantiles of the standard distribution back into a nonlinear scale of plotting positions. There is no difference between the two versions except for the horizontal scale. With probability paper, the horizontal axis can be directly interpreted as the percent probability of occurrence, the plotting position times 100 (Figure 9.1).

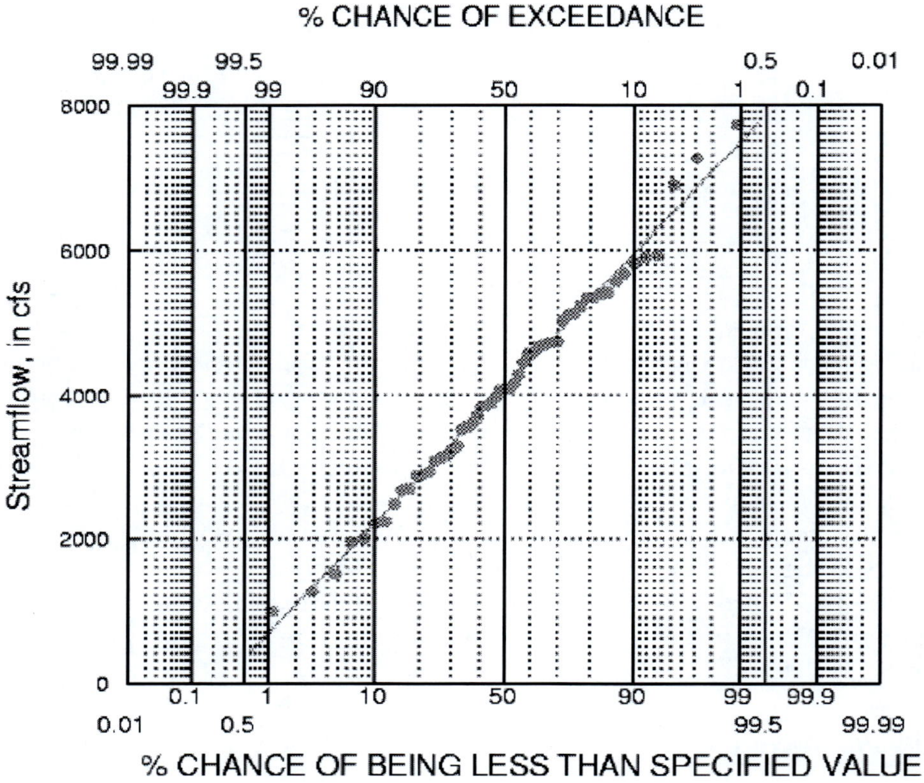

Figure 9.1. Schematic of probability plot in probability paper.

9.2. Correlation and Regression

9.2.1. Correlation

Statistical correlation is a statistical technique which tells us if two variables are related.

For example, consider that price and demand of a commodity are related variables. When price increases, demand will tend to decreases and vice versa. If the change in one variable is accompanied by a change in the other, then the variables are said to be correlated. We can therefore say that price and demand are correlated.

Partial Correlation Analysis

Partial correlation analysis involves studying the linear relationship between two variables after excluding the effect of one or more independent factors. For example, study of partial correlation between price and demand would involve studying the relationship between price and demand excluding the effect of money supply, exports, etc.

Multiple Correlation Coefficient

The Sample Multiple Correlation Coefficient, R, is a measure of the strength of the association between the independent (explanatory) variables and the one dependent (prediction) variable.

Here, we study the effects of all the independent variables simultaneously on a dependent variable. For example, the correlation co-efficient between the yield of paddy (X_1) and the other variables, viz. type of seedlings (X_2), manure (X_3), rainfall (X_4), humidity (X_5) is the multiple correlation co-efficient $R_{1.2345}$. This co-efficient takes value between 0 and +1.

It can also be interpreted as the proportion of the variance of the dependent variable explained by the independent variables. When the independent variables (used for predicting the dependent variable) are pairwise orthogonal, the multiple correlation coefficient is equal to the sum of the squared coefficients of correlation between each independent variable and the dependent variable. This relation does not hold when the independent variables are not orthogonal. The significance of a multiple coefficient of correlation can be assessed with an F-ratio.

Interpretation of R

- Strength of the Association: The strength of the association is measured by the sample Multiple Correlation Coefficient,R. Rcan be any value from 0 to +1.
    - The closer *R* is to one, the stronger the linear association is.
    - If Requals zero, then there is no linear association between the dependent variable and the independent variables.

Unlike the simple correlation coefficient, *r*, which tells both the strength and direction of the association, Rtells only the strength of the association. Ris never a negative value. This can be seen from the formula below, since the square root of this value indicates the positive root.

Formula for R

Formula for two independent variables, X_1 and X_2:

$$R = \frac{\sqrt{r_{yx_1}^2 + r_{yx_2}^2 - 2r_{yx_1}(r_{yx_2})(r_{x_1x_2})}}{\sqrt{1 - r_{x_1x_2}^2}}$$

where,

r_{yx1} = correlation coefficient between y and x_1

r_{yx2} = correlation coefficient between y and x_2

r_{x1x2} = correlation coefficient between x_1 and x_2

General Procedure

There are really two steps to compute R:

- Compute each of the paired simple correlation coefficients.
- Substitute these values in the formula for R.

In Excel

Procedure

1. From the main menu select **Tools>Data Analysis**
2. From the Data Analysis dialog box, scroll down and select **Regression**. Click **OK**
3. The Regression dialog box should appear.
   - Put your cursor in the field for the **Input Y Range**.
   - Click the **Collapse Dialog button** (the button at the right end of the field. It has a small red arrow on the button. This will Collapse the dialog box and take you to the spreadsheet.
   - Highlight the range for the Water Consumption including the title. (i.e., Cells C1-C8)
   - Press the **Return to Dialog button**. (The button at the end of the field, with the small red arrow on it.)
   - Click in the field for the **Input X Range**.
   - Click the **Collapse Dialog button.**
   - Highlight the range for the **Temperature and Time Mowing** including the title. (i.e., **Cells A1-B8)**
   - Note: You have to enter temperature and time mowing in the X range.
   - Press the **Return to Dialog button**.
   - Mark the radio button for **New Worksheet**. Click in the field box and type the word **multicorrelation**.
4. Save your work. Select **FILE > SAVE AS**. Call the data set ***Excel multi Data***.

9.2.2. Regression Analysis

Multiple Regression Analysis

Multiple linear regression (MLR) is the extension of simple linear regression (SLR) to the case of multiple explanatory variables. The goal of this relationship is to explain as much as possible of the variation observed in the response (y) variable, leaving as little variation as possible to unexplained "noise."

Multiple regression analysis is a powerful technique used for predicting the unknown value of a variable from the known value of two or more variables- also called the predictors.

More precisely, multiple regression analysis helps us to predict the value of Y for given values of $X_1, X_2, \ldots, X_k$.

For example, the yield of rice per acre depends upon quality of seed, fertility of soil, fertilizer used, temperature, rainfall. If one is interested to study the joint affect of all these variables on rice yield, one can use this technique. An additional advantage of this technique is that it also enables us to study the individual influence of these variables on yield.

When are multiple explanatory variables required? The most common situation is when scientific knowledge and experience tells us they are likely to be useful. For example, average runoff from a variety of mountainous basins is likely to be a function both of average rainfall and of altitude; average dissolved solids yields are likely to be a function of average rainfall, percent of basin in certain rock types, and perhaps basin population. Concentrations of contaminants in shallow groundwater are likely to be functions of both source terms (application rates of fertilizers or pesticides) and subsurface conditions (soil permeability, depth to groundwater, etc.).

The use of MLR might also be indicated by the residuals from a simple linear regression (SLR). Residuals may indicate there is a temporal trend (suggesting time as an additional explanatory variable), a spatial trend (suggesting spatial coordinates as explanatory variables), or seasonality (suggesting variables which indicate which season the data point was collected in). Analysis of a residuals plot may also show that patterns of residuals occur as a function of some categorical grouping representing a special condition such as: on the rising limb of a hydrograph, at cultivating time, during or after frontal storms, in wells with PVC casing, measurements taken before 10:00 a.m., etc. These special cases will only be revealed by plotting residuals versus a variety of variables - in a scatter plot if the variable is continuous, in grouped boxplots if the variable is categorical. Seeing these relationships should lead to definition of an appropriate explanatory variable and its inclusion in the model if it significantly improves the fit.

The Multiple Regression Model

In general, the multiple regression equation of Y on $X_1, X_2, \ldots, X_k$ is given by:

$$Y = b_0 + b_1 X_1 + b_2 X_2 + \ldots\ldots\ldots\ldots\ldots\ldots\ldots + b_k X_k + \epsilon$$

where:

y is the response variable
b0 = intercept
b_1 = slope coefficient for the first explanatory variable
b_2 = slope coefficient for the second explanatory variable
b_k = slope coefficient for the kth explanatory variable, and
€ = remaining unexplained noise in the data (the error).

There are k explanatory variables, some of which may be related or correlated to each other. It is therefore best to avoid calling these “independent” variables. They may or may not be independent of each other. Calling them explanatory variables describes their purpose: to explain the variation in the response variable.

Thus if $b_i = 2.5$, it would indicates that Y will increase by 2.5 units if X_i increased by 1 unit.

Assumptions

Multiple regression technique does not test whether data are linear. On the contrary, it proceeds by assuming that the relationship between the Y and each of X_i’s is linear. Hence as a rule, it is prudent to always look at the scatter plots of (Y, X_i), i = 1, 2, …, k. If any plot suggests non linearity, one may use a suitable transformation to attain linearity.

Another important assumption is non existence of multicollinearity- the independent variables are not related among themselves. At a very basic level, this can be tested by computing the correlation coefficient between each pair of independent variables.

Other assumptions include those of homoscedasticity and normality.

Multiple regression analysis is used when one is interested in predicting a continuous dependent variable from a number of independent variables. If dependent variable is dichotomous, then logistic regression should be used.

Significance Test

In order to assess the significance of a given $R^2_{Y.1,...,J}$, we can compute an F-ratio as:

$$F = \frac{R^2_{Y.1...j}}{1 - R^2_{Y.1...j}} \times \frac{N-J-1}{J}$$

Under the usual assumptions of normality of the error and of independence of the error and the scores, this F-ratio is distributed under the null hypothesis as a Fisher distribution with $N_1 = J$ and $N_2 = N\text{-}J\text{-}1$ degrees of freedom.

Dealing with Seasonality

There are many instances where changes between different seasons of the year are a major source of variation in the Y variable. As with other exogenous effects, seasonal variation must be compensated for or "removed" in order to better discern the trend in Y over time. If not, little power may be available to detect trends which are truly present. We may also be interested in modeling the seasonality to allow different predictions of Y for differing seasons.

Most concentrations in surface waters show strong seasonal patterns. Streamflow itself almost always varies greatly between seasons. This arises from seasonal variations in precipitation volume, and in temperature which in turn affects precipitation type (rain versus snow) and the rate of evapotranspiration. Some of the observed seasonal variation in water quality may be explained by accounting for this seasonal variation in discharge. However, seasonality often remains even after discharge effects have been removed (Hirsch et al. 1982). Possible additional causes of seasonal patterns include biological activity, both natural and managed activities such as agriculture. For example, nutrient concentrations vary with seasonal application of fertilizers and the natural pattern of uptake and release by plants. Other effects are due to different sources of water dominant at different times of the year, such as snow melt versus intense rainfall. Seasonal rise and fall of ground water can also influence water quality. A given discharge in one season may derive mostly from ground water while the same discharge during the another season may result from surface runoff or quick flow through shallow soil horizons. The chemistry and sediment content of these sources may be quite different.

Techniques for dealing with seasonality fall into three major categories (Helsel and Hirsch, 2002). One is fully nonparametric, one is a mixed procedure, and the last is fully parametric. In the first two procedures, it is necessary to define a "season." In general, seasons should be just long enough so that there is some data available for most of the seasons in most of the years of record. For example, if the data are primarily collected at a monthly frequency, the seasons should be defined to be the 12 months. If the data are collected quarterly then there should be 4 seasons, etc.

Polynomial Regression

Sometimes data fits better with a polynomial curve. A situation may arise where the linear relationship between two variables is inadequate. For example, yield of a crop usually increases with the increase in rainfall up to a certain level above which the yield may decrease. So, here the linear representation is not adequate.

$$\text{quadratic model: } y = \beta_0 + \beta_1 x + \beta_2 x^2 + \epsilon$$

$$\text{higher order polynomial: } y = \beta_0 + \beta_1 x + \beta_2 x^2 + \beta_3 x^3 + \ldots + \epsilon$$

9.3. Data Transformation

Transformations are used for three purposes:

1. To make data more symmetric,
2. To make data more linear, and
3. To make data more constant in variance.

Serious problems can occur when procedures assuming symmetry, linearity, or homoscedasticity (constant variance) are used on data which do not possess these required characteristics. Transformations can produce these characteristics, and thus the use of transformed variables meets an objective. Employment of a transformation is not merely an arbitrary choice.

When dealing with several similar data sets, it is probably better to find one transformation which works reasonably well for all, rather than using slightly different ones for each. It must be remembered that each data set is a sample from a larger population, and another sample from the same population will likely indicate a slightly different 'best' transformation (USGS, 2002).

9.4. Nonparametric Statistics

Nonparametric statistics are those data that do not assume a prior distribution. When an experiment is performed or data collected for some purpose, it is usually assumed that it fits some given probability distribution, typically the normal distribution. This is the basis on which the data is interpreted. When these assumptions are not made, it becomes nonparametric statistics.

There are several advantages of using nonparametric statistics:

As can be expected, since there are fewer assumptions that are made about the sample being studied, nonparametric statistics are usually wider in scope as compared to parametric statistics that actually assume a distribution. This is mainly the case when we do not know a lot about the sample we are studying and making a priori assumptions about data distributions might not give us accurate results and interpretations. This directly translates into an increase in robustness.

However, there are also some disadvantages of nonparametric statistics. The main disadvantage is that the degree of confidence is usually lower for these types of studies. This means for the same sample under consideration, the results obtained from nonparametric statistics have a lower degree of confidence than if the results were obtained using parametric statistics. Of course, this is assuming that the study is such that it is valid to assume a distribution for the sample.

There are many experimental scenarios in which we can assume a normal distribution. For example if an experiment looks at the correlation between a healthy morning breakfast and IQ, the experimenter can assume beforehand that the IQs of the sample size follow a normal distribution within the sample, assuming the sample is chosen randomly from the population. On the other hand, if this assumption is not made, then the experimenter is following nonparametric statistics methods.

However, there could be another experiment that measures the resistance of the human body to a strain of bacteria. In such a case, it is not possible to determine if the data will be normally distributed. It might happen that all people are resistant to the strain of bacteria under study or perhaps no one is. Again, there could be other considerations as well. It could be that people of a particular ethnicity are born with that resistance while none of the others are. In such cases, it is not right to assume a normal distribution of data. These are the situations in which nonparametric statistics should be used. There are many tests that tell us whether the data can be assumed to be normally distributed or not.

Nonparametric tests are sometimes called *distribution free* statistics because they do not require that the data fit a normal distribution. More generally, nonparametric tests require less restrictive assumptions about the data. Another important reason for using these tests is that they allow for the analysis of categorical as well as rank data.

Since nonparametric tests require fewer assumptions and can be used with a broader range of data types, the question becomes, "Why not use them all of the time?" Parametric tests are often preferred because:

1. They are *robust*.
2. They have greater *power efficiency*, in other words, they have greater *power* relative to the sample size.
3. They provide unique information (e.g., the interaction in a factorial design).
4. Parametric and nonparametric tests often address two different types of questions.

It is important to note that even with metric data, if assumptions are badly violated, nonparametric tests are likely to be employed.

9.5. Trend Analysis of Hydro-Meteorological Data

9.5.1. Conceptualization

Climate change is consistently associated with changes in a number of components of the hydrological cycle and hydrological systems such as precipitation patterns and intensity, and extreme weather events. In order to improve our understanding of precipitation and provide advice for water resource management under climate change, many researchers have conducted trend and variability analyses of hydrological time series. Lets have some practical example where we need trend analysis.

(a) Groundwater levels were recorded for many wells in a study area over 20 years. During the ninth year, development of the area increased withdrawals dramatically. Is there evidence of decreasing water levels in the region's wells after versus before the increased pumpage?
(b) Climate change issue is a concern now-a-days. Is the climatic elements (e.g., temperature (maximum/minimum), yearly/seasonal/monthly rainfall, humidity, evaporation, etc.) changing significantly over time?
(c) Over the years (25 yrs), the fluctuation of rainfall is suspected to decrease the yield of rainfed cereal productions. Is the decreasing rainfall decreased yield of cereals?
(d) Concentrations and loads of phosphorus have been observed at numerous tributaries to an important estuary over a 20-year period. Have concentrations and/or loads changed over time? Have concentrations changed when changing flow conditions are taken into account (the early years were during a very dry period), or are all changes simply due to more precipitation in the latter years? Is there an observable effect associated with a ban on phosphorus compounds in detergents which was implemented in the middle of the period of record?

Trend analysis is one of common parts for detecting changes in hydrological time series, which has received considerable attention in the recent years. Many important variables relating to hydrological changes, e.g., the trend of natural runoff, streamflow, and snowmelt runoff in some rivers, have been widely researched. Rainfall trend analysis has been used by climatologists and hydrologists for many years, and is frequently used nowadays to trace the impact of climatic variability and change due to natural and anthropogenic influences across the globe. The accompanying variability in available surface and subsurface water resources is associated with global and regional climatic variability governed by complex land–atmospheric–ocean processes.

9.5.2. Procedure for Trend Analysis

The explanatory variable of interest is usually time, though spatial or directional trends (such as downstream order or distance downdip) may also be investigated. Tests for trend have been of keen interest in environmental sciences over the last 10-15 years. Detection of both sudden and gradual trends over time with and without adjustment for the effects of confounding variables can been employed.

The magnitude of the trend in the time series can be determined using Sen's estimator (Sen, 1968). This method has been widely used for determining the magnitude of trend in hydro-meteorological time series. The statistical significance of the trend in monthly, seasonal and annual series can be analysed using the non-parametric Mann-Kendall (MK) test (Mann, 1945; Kendall, 1975). The MK test checks the null hypothesis of no trend versus the alternative hypothesis of the existence of increasing or decreasing trend. No pre-whitening of the data series is necessary If the sample size is large (n > 50) and the slope of trend is high (>0.01).

Mann-Kendall test

The Mann-Kendall (Mann 1945, Kendall 1975) trend test is a rank-based non-parametric test originally developed for non-correlated data. The method has been further developed by some researchers (Kendall, 1975; Sneyers, 1990). Not only can it detect any possible trend in temperature or hydrological series, but it can also test whether such trends are statistically significant or not.

To perform the Mann-Kendall test, Kendall's S_k is calculated first from all pairs (x_i, x_j) observations of a data series X with length n as follows:

$$S_k = \sum_{i=1}^{k} r_i \quad (k = 2, 3, \ldots, n)$$

Where,

$$r_i = \begin{cases} 1 \; if xi > xj \\ 0 \; if xi \leq xj \end{cases} \; (j = 1, 2, \ldots, i)$$

The test is based on the statistic τ defined as:

$$\tau = \frac{4 \sum S_k}{n(n-1)} - 1$$

Under the null hypothesis that there is no trend, the statistic τ is normally distributed with zero mean and variance:

$$var(\tau) = \sigma_\tau^2 = \frac{2(2n+5)}{9n(n-1)}$$

The significance test is based upon the formula:

$$Z = \pm\tau \sqrt{\frac{9n(n-1)}{2(2n+5)}}$$

Here, the Z converges rapidly to a standard normal distribution as *n* increases. At a specified level of significance of α, a standard Z_α value can be obtained from a table of standard normal distribution, with $Z_{\alpha/2}$ being the critical value.

The value of Z_t is greater than the critical value $Z_{\alpha/2}$ (i.e., $Z_t > Z_{\alpha/2}$) in the case of a positive trend, and greater than $Z_{-\alpha/2}$ in the case of a negative trend.

Relevant Journals

The Statistician
J. Am. Stat. Assoc.
Econometrica
Environmental and Ecological Statistics
Journal of Hydrology
Hydrological Sciences Journal

Questions

1. What are the methods of graphical presentation of data?
2. Discuss the condition and assumptions for: simple correlation, multiple correlation, simple regression, multiple regression
3. What are the merits of using non-parametric statistics?
4. Briefly narrate the method of trend analysis of hydrological data

REFERENCES

Blom, G. (1958). Statistical Estimates and Transformed Beta Variables: John Wiley, New York, 68-75, 143-146.

Chambers, J. M., Cleveland, W. S., Kleiner, B. & Tukey, P. A. (1983). Graphical Methods for Data Analysis: PWS-Kent Publishing Co., Boston, 395 p.

Cunnane, C. (1978). Unbiased plotting positions - a review: *J. of Hydrology*, *37*, 205-222.

Gringorten, I. I. (1963). A plotting rule for extreme probability paper: *J. of Geophysical Research*, *68*, 813-814.

Hazen, A. (1914). Storage to be provided in the impounding reservoirs for municipal water supply: *Trans. Am. Soc. of Civil Engineers*, *77*, 1547-1550.

Helsel, D. R. & Hirsch, R. M. (2002). Statistical Methods. Chapter A3, Techniques of Water-Resources Investigations of the United States Geological Survey. Book 4, Hydrologic Analysis and Interpretation. U.S. Department of The Interior, and U.S. Geological Survey.

Hirsch, R. M., Slack, J. R. & Smith, R. A. (1982). Techniques of trend analysis for monthly water quality data: *Water Resources Research*, *18*, 107-121.

Iman, R. L. & Conover, W. J. (1983). A Modern Approach to Statistics: John Wiley and Sons, New York, 497 p.

Kendall, M. G. (1975). Rank Correlation Methods, 4th edition: Charles Griffin, London. 202 p.

Looney, S. W. & Gulledge, T. R. (1985b). Probability plotting positions and goodness of fit for the normal distribution: *The Statistician*, *34*, 297-303.

Mann, H. B. (1945). Nonparametric test against trend: *Econometrica*, *13*, 245-259.

Sen, P. K. (1968). Estimates of regression coefficient based on Kendall's tau: *J. Am. Stat. Assoc.*, *63*, 1379-1389.

Sneyers, R. (1975). Sur l'analyse statistique des series d'observations. Geneva, Switzerland: World Meteorological Organization, Technical Note 143.

Weibull, W. (1939). The Phenomenon of Rupture in Solids: Ingeniors Vetenskaps Akademien Handlinga 153, Stockholm, p. 17.

Chapter 10

MODELS IN WATER RESOURCES MANAGEMENT

ABSTRACT

Model is a simplified representation of reality. It may be a system of equations that shows how parameters and variables of interest are related to one another. Model is ideal to integrate the complex system and to obtain the answer for unknown situation. It is a useful tool in obtaining answers in the choice of a decision or policy. A model can be of many shapes, sizes, and styles. Models of different types are discussed in this chapter. Different issues of modeling aspects such as calibration, sensitivity analysis, validation, etc. have been described. Different up-to-date statistical indicators for model performance evaluation have been discussed, and overview of some of the commonly used models is included.

10.1. BACKGROUND/NEED OF A MODEL

A model is merely a useful tool in obtaining answers in the choice of a decision or policy. Some of the fundamental problems of water resources management planning are how to obtain knowledge of specific process within a complex system of interacting and interdependent phenomena, and then how to reintegrate such knowledge so as to obtain a comprehensive and accurate solution of the phenomena. Model is ideal to integrate the complex system and to obtain the answer if the condition is. Scenario simulations are needed for the purpose of assisting managers/planners in management decisions, predicting impacts of land and/or water use decisions, and predicting the consequences of probable climatic variations.

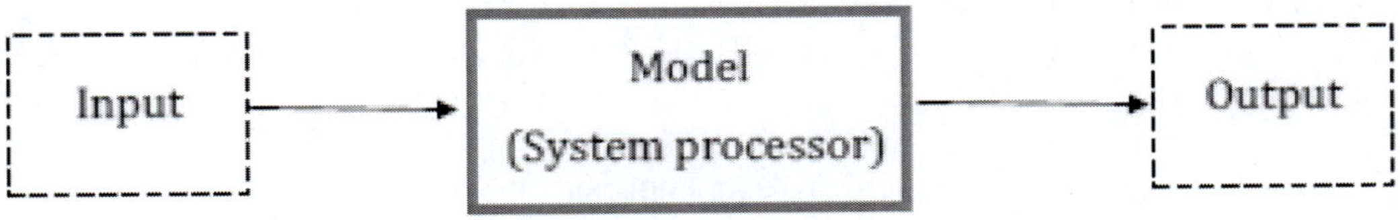

Figure 10.1. Schematic representation of a model.

10.2. Basics of Model: General Concepts, Types, Formulation and Evaluation System

10.2.1. General Concepts

A model is not the real world but only an approximation of reality to help us better understand real world systems. It is a tool to obtain promising water management strategies. In general all models have an interface, information input, an information processor, and an output of expected results (**Figure10.1**).

10.2.2. Types of Model

Models can be broadly classified into two major groups:

- Physically based *or* process based models, and
- Empirical models

Physically Based or Process Based Model

In this category, the processes are conceptually represented using mathematical equations. A process based model normally conceptualizes represents all important physical processes occurring during an event as well as between events (where applicable). Such a model is derived from theoretical approaches to causal relationship. This type of models are accurate over a wide range of conditions.

Empirical Models (or Black Box Models)

This is mathematical model, and inputs and outputs are related by empirical equations. The internal mechanism of the process (or causal relationship) is not explicitly described (or unknown, i.e., like a 'black box'). In this category, empirically derived formulas (relies upon experiments and observations) are used to predict the required output. The empirical data are derived from a limited sample of projects.

10.2.3. Some Relevant Terminologies

Simulation

Simulation reproduces the essence of a system without reproducing the system itself. Mathematical simulation is a technique whereby the behavior of a system is described using a combination of mathematical and logical relationships. A system is a part of reality that the engineers want to study.

Model parameter: An item that is constant in the model; a number.

Variable: An item of interest that can change value depending on the other components of the model.

Constraint: Any condition that places bounds on the solution is a constraint.

Initial Condition

Values of the state and control variables at the beginning of the time period of interest. The behaviors of all dynamic system are dependent upon their initial conditions. In classical mechanics, the initial conditions of systems are known.

Boundary Condition

Boundary conditions are the set of conditions specified for the behavior of the solution to a set of differential equations at the boundary of its domain. Boundary conditions are important in determining the mathematical solutions to many physical problems.

Objective Function

It is an equation that measures how well the system is attaining an objective defined by the modeler. The objective function is expressed in terms of the state variables, control variables, and parameters of the system.

Dynamic System

A system that contains time as one of the variables. The changes of parameters and variables and inter-relationship between them occur through time.

Analytical solution: A mathematical solution to the objective that is based on an analytical analysis. Analytical solutions are more general than numerical results.

Sensitivity Analysis

It is the study of how variation (uncertainty) in the output of a mathematical model can be apportioned quantitatively, or qualitatively, to different sources of variation in the input of a model. In models involving many input variables, sensitivity analysis is an essential part of model building and quality assurance.

The sensitivity analysis of a model is the identification of the most important parameters with respect to the results of the model, and to decrease the number of parameters used in the calibration process.

Sensitivity analysis is performed by changing a particular variable or parameter while keeping all other variables or parameters constant, and observed how the output is changed. It is an important method for checking the quality of a given model, as well as a powerful tool for checking the robustness and reliability of its analysis.

Verification/validation: The process of determining that a model implementation accurately represents the developer's conceptual description of the model and the solution to the model.

Model evaluation: It is the testing of a model's capability to simulate the result.

Confidence: Probability that a numerical estimate will lie within a specified range.

Uncertainty: A potential deficiency in any phase or activity of the modeling or experimentation process that is due to inherent variability or lack of knowledge.

Implicit and Explicit Method

When a direct computation of the dependent variables can be made in terms of known quantities, the computation is said to be explicit. One can compute values on new level (time or space) by explicit formula.

When the dependent variables are defined by coupled sets of equations, and either a matrix or iterative technique is needed to obtain the solution, the numerical method is said to be implicit. To find a value on new level, one must solve the linear system of equations.

Finite Element and Finite Difference Method

The finite element method is a method through which any continuous function can be approximated by a discrete model which consists of a set of values of the given function at a finite number of pre-selected points in its domain, together with piecewise approximation of the function over a finite number of connected disjunct sub-domains. These sub-domains are called finite elements and are determined by the pre-selected points. The pre-selected points are called nodal points or nodes.

Finite difference methods are numerical methods for approximating the solutions to differential equations using finite different scheme. It consists of replacing (transforming) the partial derivative expression with approximately equivalent difference quotients (equations) over a small interval. That is, it transforms the continuous domain of the state variables by a network or mesh of discrete points.

10.2.4. Considerations in Model Development

Model development begins with a problem to be solved, or a question or hypothesis. At that stage, a decision must be made about the type of model the researcher will develop. For all cases, developing a model may not necessary. Simply developing a clear conceptual model can answer to a question. However, many complex problems (having complex interactions among different processes) can not be solved with conceptual model alone. The process of developing a mathematical model can be extremely helpful in clarifying the relevant processes. It directs or forces the modeler to think clearly about the important processes and relationships that govern important features of the problem.

The key features that should be considered in model development are:

- Simplifying assumptions should be made;
- Initial and boundary conditions should be identified;
- The range of applicability of the model should be understood.

Formulation of Model

Modeling process should begin with a definition of goal(s). The goal may be find from the answer of the question – 'What will be the model be used for – to solve a specific problem, to learn more about the system, to evaluate alternative management strategies, etc.? The answer will help in determining the structure of the model, the data required, and the expected output.

The next step is a conceptualization of the prototype, including selection of scales of time and space, dimensionality and discretization. This is followed by formulation as a set of differential equations, statistical properties, empirical expressions, or combinations of these. Computational representation of model formulation implicitly requires selection of a method of solution.

Once the model is calibrated with a specific set of parameter values, it may be tested again against another set of prototype data in a process of *verification* to ascertain whether the degree of conformity with the prototype under the new conditions is consistent with that of calibration. If the model fails this test, recalibration may be necessary. Calibration/verification is sometimes considered a single process in model development, termed as *validation*.

Sensitivity testing is an integral part of the model development process, usually accompanying the calibration/verification phases. It provides information on the relative importance of model parameters and their influence on model response in terms of state variables simulated. Within the period of model development, specially at the time of sensitivity analysis, it is appropriate to fix the character of the model, i.e., to provide documentation for the user.

The real test of model performance is in application. This may reveal deficiencies in model, requiring appropriate modification in structure, formulation, solution technique, parameter estimates, or basic data.

10.2.5. Model Calibration, Validation and Evaluation

The model evaluation procedure involves both calibration and validation of the model. Before simulation models are used, they need to be calibrated and validated for the conditions under which they will be used.

Model Calibration

The calibration process involves the adjustment of selected model parameters within an expected range so as to minimize the discrepancies between observed and predicted values. Calibration of the model is typically accomplished by adjusting input parameters and rate constants so that output values are as close as possible to measured values in the calibration data set.

This approach to calibrating a model has the disadvantage of requiring a calibration data set. In addition, the approach can be very time consuming, especially if the range of values explored for each parameter and rate constant is relatively large. Furthermore, since the process adjusts several input parameters and rate constants at the same time, inadequacies in the modeling of one process may be coincidentally compensated for when the input parameter or rate constant is changed for another process. Consequently, this approach may lead to good simulation for the calibration data set, but will not necessarily guarantee good simulations for independent data sets.

Another approach to calibration is to conduct laboratory studies to determine the rate constants and input parameters for each of the important processes in the model, and then use those rate constants and parameters to validate the model.

Model Verification/Validation

Validation is essentially an independent test of the model, where the model predictions are compared with data not used in the calibration testing. The goal of verification is to quantify confidence in the predictive capacity of the model by comparison with experimental data.

In contrast to traditional experiments, validation experiments are performed to generate high quality data for the purpose of assessing the accuracy of a model prediction. To qualify as a validation test, the specimen geometry, initial conditions, boundary conditions, and all other model input parameters must be prescribed accurately.

Evaluation of model performance should include both statistical criteria and graphical display. A model is a good representation of reality only if it can predict an observable phenomenon with acceptable accuracy and precision.

10.2.6. Statistical Indicators for Model Performance Evaluation

One of the most common issues raised concerning model usage is that of reliability. There is a pressing need to assess model validity in quantitative terms, e.g., statistical limits of model performance against the prototype system. The following statistics are generally used to indicate overall model performance: Bias, Mean absolute bias, root mean square error, relative error, model efficiency, Index of Agreement, Coefficient of residual mass, Coefficient of determination (Fox, 1981; Willmott et al., 1985; Willmott etal., 2011; Loague et al., 1989; Loague and Green, 1991; Retta et al., 1996; Ali and Abustan, 2014). Ali and Abustan (2014) reviewed the performance of indicators, identified their demerits/shortcomings, and suggested a new indicator,

(i) Bias (i.e., error) or Mean Bias:

$$\text{Bias} = \frac{1}{N}\sum_{i=1}^{N}(S_i - M_i) \tag{10.1}$$

where S and M are the simulated (or predicted) and measured (or observed) values for the ith observation and N is the number of observations.

(ii) Mean absolute bias:

$$MAE = \frac{1}{N}\sum_{i=1}^{i=N} |S_i - M_i| \tag{10.2}$$

(iii) Root mean square error (RMSE), which quantifies the dispersion between simulated and measured (or observed) data:

$$RMSE = \sqrt{\frac{1}{N}\sum_{i=1}^{N}(S_i - M_i)^2} \tag{10.3}$$

Ideally, the value of RMSE should be zero.

(iv) Relative error (RE):

$$RE = \frac{RMSE}{\bar{y}}.100 \quad (10.4)$$

where $\bar{y}$ is the mean of observed values.

(v) Model efficiency (EF): Model efficiency is calculated as:

$$EF = \frac{\sum (M-\bar{M})^2 - \sum (S-M)^2}{\sum (M-\bar{M})^2} \quad (10.5)$$

Where,
$\bar{M}$ = measured mean
An ideal value of *EF* is unity.
(vi) Index of Agreement (IA):

$$IA = 1 - \frac{\sum_{i=1}^{N} (O_i - S_i)^2}{\sum_{i=1}^{N} (O_i' + S_i')^2}, \; 0 \le \text{IA} \le 1 \quad (10.6)$$

where $O_i' = |O_i - \bar{S}|$, $S_i' = |S_i - \bar{S}|$, O_i is the observed value, S_i is the simulated value, $\bar{S}$ is the simulated mean.

(vii) **Coefficient of residual mass (CRM)**:
It is expressed as:

$$CRM = \frac{(\sum O_i - \sum S_i)}{\sum O_i} \quad (10.7)$$

(viii) **Coefficient of determination (CD)**:
It is expressed as:

$$CD = \frac{\sum (O_i - \bar{O})^2}{\sum (S_i - \bar{O})^2} \quad (10.8)$$

(ix) **Percent mean absolute relative error (PMARE)**
It is the 'mean absolute relative error,' expressed in percentage.

$$PMARE(\%) = \frac{100}{n} \sum_{i=1}^{n} \frac{Abs(O_i - P_i)}{O_i} \quad (10.9)$$

Where, "*Abs*' indicates 'absolute value (of the difference between observed and simulated value). Theoretically, the value of PMARE ranges from 0% to ∞ (positive infinity). Ali and Abustan (2013) suggested practical/ acceptable range of 0<PMARE<100.

10.3. Overview of Some of the Commonly Used Models

Models used in different sectors of water resources are:

10.3.1. Irrigation management and planning: CROPWAT, UPFLOW, AquaCrop, CropSyst
10.3.2. Surface water hydrology/Storm water management: SWMM, SWAT, HEC-ARS, WEAP
10.3.3. Groundwater movement/hydrology: HYDRUS1D, HYDRUS2D,
10.3.4. Coupled surface water – groundwater model

10.3.1. Model for Irrigation Planning and Decision Support System

10.3.1.1. CROPWAT Model

The FAO developed CROPWAT software (Clarke, 1998) to calculate ET_0 and irrigation planning. The ET_0estimation is based on Penman-Monteith (P-M) equation. The inputs for ET0 estimation are: (i) maximum temperature, (ii) minimum temperature, (iii) Average humidity, (iv) wind speed, (v) sunshine hour, (vi) latitude, and (vii) altitude of the location.

10.3.1.2. UPFLOW - Model for Upward Flux Estimation

UPFLOW is a software tool developed to estimate the expected upward flow from a shallow water table in a given soil profile and to evaluate the effects of environmental conditions on the upward flow (Raes, 2002; Raes and Deproost, 2003).

The input in the UPFLOW model are: average evapotranspiration (ET) during the time period, expected mean soil water content, average depth of watertable (WT) below the soil surface, crop type, the characteristics of the various layers of the soil profile and their thickness, and the salt content of the groundwater (if groundwater contains salts).

For the given environmental conditions, UPFLOW displays the expected steady upward flow [mm/day] from the water table to the topsoil, the soil water content [vol%] expected in the topsoil, the amount of salt transported upward during the given period [t/ha-year, if the water table contains salts,], the degree of water logging [%] in the root zone (if any), and a graphical display of the soil water profile above the water table.

In the model, the steady upward flow to the topsoil is estimated according to De Laat (1980, 1995):

$$z = \int_0^h \frac{K(h)}{q + K(h)} dh$$

Where z(m) is the vertical co-ordinate, q the constant upward flux ($m^3.m^{-2}.d^{-1}$) of water, h the soil matric potential per unit weight of water (m), and K(h) the hydraulic conductivity (m/d).

Given the K-h and θ –h relation for the various soil layers of the profile above the water table, UPFLOW is able to determine the maximum flux that can flow to the top soil by checking that the simulated soil water content (derived from the moisture profile) remains below the specified mean water content in the top soil. UPFLOW calculates the amount of water that the plant roots extract according to Feddes et al. (1978). Since the water flow inside the soil profile is assumed to be steady, the capillary rise from the water table to the topsoil can never exceeds the ET demand of the atmosphere. Mean soil evaporation or crop evapotranspiration for a given period from climatic, soil and crop parameters are calculated in the model according to Allen et al. (1998).

10.3.1.3. AquaCrop Model

The AquaCrop model is a water-driven simulation model to simulate the yield of field crops. It was developed by the Land and Water Division of FAO. AquaCrop predicts biomass production through simulation of crop green foliage. From biomass, grain yield is predicted using harvest index. Detail principles, methods, and capabilities of AquaCrop can be found in Raes et al. (2010) and Steduto et al. (2009).

It simulates yield response to water of herbaceous crops, and is particularly suited to address conditions where water is a key limiting factor in crop production. AquaCrop attempts to balance accuracy, simplicity, and robustness. It uses a relatively small number of explicit and mostly-intuitive parameters and input variables requiring simple methods for their determination.

AquaCrop has a simple, user-friendly structure, and employs 33 crop input parameters that can be observed easily in the fi eld; for example, the percentage of canopy cover instead of leaf area index (LAI) and other biomass-related physiological inputs; numerical and/ or descriptive characterization of crop water stress tolerance, soil texture, and nutrient input. The model maintains a signifi cant number of main output data, including the simulation of canopy cover, biomass and soil water components over the whole growing cycle, and the final harvestable yield.

Applications of AquaCrop include:

- assessing water-limited, attainable crop yields at a given geographical location
- as a benchmarking tool, comparing the attainable yields against actual yields of a field, farm, or region, to identify the yield gap and the constraints limiting crop production
- assessing rainfed crop production on the long term
- developing irrigation schedules for maximum production (seasonal strategies and operational decision-making), and for different climate scenarios
- scheduling deficit and supplemental irrigation
- evaluating the impact of fixed delivery irrigation schedules on attainable yields
- simulating crop sequences
- carrying out future climate scenario analyses

- optimizing a limited amount of water available (economic, equitability, and sustainability criteria)
- evaluating the impact of low fertility and of water-fertility interactions on yields
- assessing actual water productivity (biological and/or economic) at the field and higher scales, up to regions
- supporting decision making on water allocation and other water policy actions
- appraising the role of various water-related crop responses in yield determination for ideotype design

10.3.1.4. CropSyst Model (Cropping Systems Simulation Model)

It is a multi-year, multi-crop, daily time step crop growth simulation model (Stockle and Nelson, 1994, Stockle et al., 1994). The model simulates the soil water budget, soil-plant nitrogen budget, crop phenology, crop canopy and root growth, bio-mass production, crop yield, residue production and decomposition, soil erosion by water, and pesticide fate. The management options include: cultivar selection, crop rotation (including fallow years), irrigation, nitrogen fertilization, tillage operations, and residue management. The water budget in the model includes rainfall, irrigation, runoff, interception, water infiltration and redistribution in the soil profile, crop transpiration and evaporation.

Main Features

- Link of CropSyst simulations with GIS software is possible
- Optional runtime graphic display
- Report data graphics viewer
- Input parameter file editor
- Report format editor

Models Inputs

Four input data files are required to run CropSyst: Location, Soil, Crop, and Management files. A ‘Simulation Control file’ combines the input files as desired to produce specific simulation runs. In addition, the ‘Control file’ determines the start and ending day for the simulation, and set the values of all parameters required for initialization.

Model Output

The model outputs are yield (grain yield and biomass), water balance, nitrogen balance, residue production and decomposition, and soil erosion. There are three types of output variables:

- Daily variables computed each day.
- Yearly or annual variables provided an annual summary of values accumulated throughout the calendar year.
- Harvest variables provide harvest yield and relevant crop and soil conditions at harvest time accumulated throughout the growing season.

Output Time Step

The user may restrict the amount of information printed in the daily report by specifying a time step interval.

Limitations/Shortcomings

No specific limitations are documented.

10.3.2. Hydrologic Models

10.3.2.1. Storm Water Management Model (SWMM)

EPA's Storm Water Management Model (SWMM) is used throughout the world for planning, analysis and design related to stormwater runoff, combined and sanitary sewers, and other drainage systems in urban areas. There are many applications for drainage systems in non-urban areas as well. SWMM is a dynamic hydrology-hydraulic water quality simulation model (EPA, 2016). It is used for single event or long-term (continuous) simulation of runoff quantity and quality from primarily urban areas. The runoff component operates on a collection of sub catchment areas that receive precipitation and generate runoff and pollutant loads. The routing portion transports this runoff through a system of pipes, channels, storage/treatment devices, pumps, and regulators.

SWMM tracks the quantity and quality of runoff made within each sub catchment. It tracks the flow rate, flow depth, and quality of water in each pipe and channel during a simulation period made up of multiple time steps. SWMM 5 has recently been extended to model the hydrologic performance of specific types of low impact development (LID) controls. The LID controls that the user can choose include the following seven green infrastructure practices:

1. Permeable pavement
2. Rain gardens
3. Green roofs
4. Street planters
5. Rain barrels
6. Infiltration trenches
7. Vegetative swales

The updated model allows engineers and planners to accurately represent any combination of LID controls within a study area to determine their effectiveness in managing stormwater and combined sewer overflows.

Running under Windows, SWMM 5 provides an integrated environment for editing study area input data; running hydrologic, hydraulic and water quality simulations; and viewing the results in a variety of formats. The latter includes

- color-coded drainage area and conveyance system maps,
- time series graphs and tables,
- profile plots, and
- statistical frequency analyses.

SWMM 5 was produced in a joint development effort with CDM, Inc., a global consulting, engineering, construction, and operations firm.

Capabilities

SWMM accounts for various hydrologic processes that produce runoff from urban areas. These include

- time-varying rainfall,
- evaporation of standing surface water,
- snow accumulation and melting,
- rainfall interception from depression storage,
- infiltration of rainfall into unsaturated soil layers,
- percolation of infiltrated water into groundwater layers,
- interflow between groundwater and the drainage system,
- nonlinear reservoir routing of overland flow, and
- runoff reduction via LID controls.

Spatial variability in all of these processes is achieved by dividing a study area into a collection of smaller, homogeneous sub catchment areas. Each of the areas contains its own fraction of pervious and impervious sub-areas. Overland flow can be routed between sub-areas, between sub-catchments, or between entry points of a drainage system.

SWMM contains a flexible set of hydraulic modeling capabilities used to route runoff and external inflows through the drainage system network of pipes, channels, storage/treatment units and diversion structures. These include the ability to

- handle drainage networks of unlimited size;
- use a wide variety of standard closed and open conduit shapes as well as natural channels;
- model special elements such as storage/treatment units, flow dividers, pumps, weirs, and orifices;
- apply external flows and water quality inputs from surface runoff, groundwater interflow, rainfall-dependent infiltration/inflow, dry weather sanitary flow, and user-defined inflows;
- utilize either kinematic wave or full dynamic wave flow routing methods;
- model various flow regimes, such as backwater, surcharging, reverse flow, and surface ponding; and
- apply user-defined dynamic control rules to simulate the operation of pumps, orifice openings, and weir crest levels.

SWMM can estimate the production of pollutant loads associated with storm-water runoff. The following processes can be modeled for any number of user-defined water quality constituents:

- Dry-weather pollutant buildup over different land uses
- Pollutant wash-off from specific land uses during storm events

- Direct contribution of rainfall deposition
- Reduction in dry-weather buildup due to street cleaning
- Reduction in wash-off load due to BMPs
- Entry of dry weather sanitary flows and user-specified external inflows at any point in the drainage system
- Routing of water quality constituents through the drainage system
- Reduction in constituent concentration through treatment in storage units or by natural processes in pipes and channels

10.3.2.2. SWAT Model (Soil and Water Assessment Tool)

Development of the "Soil and Water Assessment Tool" (SWAT, Arnold et al., 1998; Neitsch et al., 2002a,b) was started in the 1990s at the United States Department of Agriculture (USDA). SWAT is a process-based and spatially semi-distributed hydrological and water quality model designed to calculate and route water, sediments and contaminants from individual drainage units (sub-basins) throughout a river basin towards its outlet. Due to this flexibility, SWAT has been used in many parts of the world. It has been used to predict the impact of management practices on water, sediment and agricultural chemical yields in large complex basins with varying soils, land use and management conditions, over long periods of time (Eckhardt et al., 2005). It has the capability of scenario generation to equip policy makers with a wider range of options. A complete description of the SWAT model can be found in Neitsch et al., (2002a,b). A short overview of the most relevant aspects related to the hydrology component has been presented below.

Within the SWAT conceptual framework, the representation of the hydrology of a basin is divided into two major parts: (a) the land phase of the hydrological cycle; and (b) the routing of runoff through the river network. For modelling the land phase, the river basin is divided in sub-basins, each one of which is composed of one or several hydrological response units (HRUs), which are areas of relatively homogeneous land use/land cover and soil types. The characteristics of the HRUs define the hydrological response of a sub-basin. For a given time step, the contributions to the discharge at each sub-basin outlet point is controlled by the HRU water balance calculations (land phase). The river network then connects the different sub-basin outlets, and the routing phase determines movement of water through this network towards internal control points, and finally towards the basin outlet (Neitsch et al., 2002a). For the land phase water balance, within SWAT evapotranspiration can be calculated using one of three methods: Penman-Monteith, Hargreaves or Priestley-Taylor. The Penman-Monteith method offers a better process description, but has high input data requirements which for practical applications will be hard to fulfil in many parts of the world. Although less physically based, the Hargreaves or Priestley-Taylor methods have the advantage of less stringent input data needs; under minimal conditions of data availability, the Hargreaves method can even be used with temperature time series as the only required measured input. For surface runoff calculations, SWAT gives the user two alternatives: (a) the use of the Soil Conservation Service curve number (SCS CN) procedure, and (b) the Green and Ampt infiltration method. For the latter method, input data at a finer-than-daily time resolution are required, whereas the CN method is lumped over time: the SCS CN approach can typically be applied using daily rainfall values. Runoff contributions from snowmelt can be incorporated by means of a temperature index, a method commonly used in water resources management applications (Walter et al., 2005).

SWAT was developed to predict the impact of land-management practices on water, sediment and agricultural chemical yields in large, complex watershed with varying soils, land-use and management conditions over long periods of time. SWAT is a continuous-time, long-term yield model with the capability of scenario generation to equip policy makers with a wider range of options.

Data Inputs for Hydrological Modeling

The SWAT model requires data on terrain, land use, soil, weather and man-made structures such as reservoirs for assessment of water-resource availability at desired locations of the drainage basin. To create a SWAT dataset, the interface needs to access Arc View map themes and database files, which provide certain types of information about the watershed.

Scenario Generation Using Hydrological Modeling

Generation of scenarios is one of the most important applications of hydrological modeling. Scenario generation helps the decision makers to analyze the potential impacts of their actions. A variety of land-use changes can be simulated and the impact on the hydrological characteristics of the basin can be analyzed. The aim of the exercise is to generate a series of scenarios or options for the stakeholders, so as to enable them to take sound rational decisions. The stakeholders, if they so desire, can use simulation modeling for some other sets of land-use changes which are more acceptable to them. Moreover, the impacts of climate changes on the hydrological regime can be effectively captured through the use of modeling techniques and the remedial actions can be taken well in time.

Scenario Generation for Climate Changes

Climate change challenges existing water-resources management practices by adding uncertainty. This adds further complications to the water-resources management. The SWAT model can be used to generate scenario for climate changes.

10.3.2.3. WEAP Model

WEAP **("Water Evaluation And Planning" system)** is a software tool that takes an integrated approach to water resources planning. WEAP is developed by the Stockholm Environment Institute's U.S. Center. It provides a comprehensive, flexible and user-friendly framework for planning and policy analysis. A detailed technical description of WEAP and its capabilities can be found in WEAP User Guide (http://www.weap21.org/). WEAP is available in many languages.

WEAP operates on the basic principle of a water balance and can be applied to municipal and agricultural systems, a single watershed or complex transboundary river basin systems. Moreover, WEAP can simulate a broad range of natural and engineered components of these systems, including rainfall runoff, baseflow, and groundwater recharge from precipitation; sectoral demand analyses; water conservation; water rights and allocation priorities, reservoir operations; hydropower generation; pollution tracking and water quality; vulnerability assessments; and ecosystem requirements. A financial analysis module also allows the user to investigate cost-benefit comparisons for projects.

The capacities of WEAP include

- ***Water balance database*****:** WEAP provides a system for maintaining water demand and supply information.
- ***Scenario generation tool*****:** WEAP simulates water demand, supply, runoff, streamflows, storage, pollution generation, treatment and discharge and instream water quality.
- ***Policy analysis tool*****:** WEAP evaluates a full range of water development and management options, and takes account of multiple and competing uses of water systems.

Steps in applications of WEAP include

- ***Study definition*****:** The time frame, spatial boundaries, system components, and configuration of the problem are established.
- ***Current accounts*****:** A snapshot of actual water demand, pollution loads, resources and supplies for the system are developed. This can be viewed as a calibration step in the development of an application.
- ***Scenarios*****:** A set of alternative assumptions about future impacts of policies, costs, and climate, for example, on water demand, supply, hydrology, and pollution can be explored. (Possible scenario opportunities are presented in the next section.)
- ***Evaluation*****:**The scenarios are evaluated with regard to water sufficiency, costs and benefits, compatibility with environmental targets, and sensitivity to uncertainty in key variables.

Main Features of WEAP

An intuitive GIS-based graphical interface provides a simple, yet powerful, means for constructing, viewing, and modifying the configuration. The user designs a schematic of the system using the mouse to "drag and drop" elements to be added to the system. These elements can be overlain on a map built from Arcview and other standard GIS and graphic files. Data for any component can be edited directly by clicking on the desired symbol in the schematic. The user may consult the context-sensitive help feature from anywhere in WEAP.

Highlights are

- Integrated water resources planning system
- Built-in models for: Rainfall runoff and infiltration, evapotranspiration, crop requirements and yields, surface water/groundwater interaction, and instream water quality
- GIS-based, graphical "drag and drop" interface
- Model-building capability with a number of built-in functions
- User-defined variables and equations
- Dynamic links to spreadsheets and other models

10.4. HEC-RAS Model

HEC-RAS is a computer program that models the hydraulics of water flow through natural rivers and other channels. The program is one-dimensional, meaning that there is no direct modeling of the hydraulic effect of cross section shape changes, bends, and other two- and three-dimensional aspects of flow. The program was developed by the US Department of Defense, Army Corps of Engineers in order to manage the rivers, harbors, and other public works under their jurisdiction; it has found wide acceptance by many others since its public release in 1995.

Features

HEC-RAS is designed to perform one-dimensional hydraulic calculations for a full network of natural and constructed channels. The following is a description of the major capabilities of HEC-RAS.

User Interface

The user interacts with HEC-RAS through a graphical user interface (GUI). The main focus in the design of the interface was to make it easy to use the software, while still maintaining a high level of efficiency for the user.

Hydraulic Analysis Components

The HEC-RAS system contains four one-dimensional river analysis components for: (1) steady flow water surface profile computations; (2) unsteady flow simulation; (3) movable boundary sediment transport computations; and (4) water quality analysis. A key element is that all four components use a common geometric data representation and common geometric and hydraulic computation routines. In addition to the four river analysis components, the system contains several hydraulic design features that can be invoked once the basic water surface profiles are computed.

Data Storage

Data Storage is accomplished through the use of "flat" files (ASCII and binary), as well as the HEC-DSS. User input data are stored in flow files under separate categories of project, plan, geometry, steady flow, unsteady flow, and sediment data. Output data is predominantly stored in separate binary files. Data can be transferred between HEC-RAS and other programs by utilizing the HEC-DSS.

Data management is accomplished through the user interface. The modeler is requested to enter a single filename for the project being developed. Once the project filename is entered, all other files are automatically created and named by the interface as needed. The interface provides for renaming, mobbing, and deletion of files on a project-by-project basis.

Reporting facilities allow for printed output of input data as well as output data. Reports can be customized as to the amount and type of information desired.

RAS Mapper

HEC-RAS has the capability to perform inundation mapping of water surface profile results directly from HEC-RAS. Using the HEC-RAS geometry and computed water surface profiles, inundation depth and floodplain boundary datasets are created through the RAS Mapper. Additional geospatial data can be generated for analysis of velocity, shear stress, stream power, ice thickness, and floodway encroachment data. In order to use the RAS Mapper for for analysis, you must have a terrain model in the binary raster floating-point format (.flt). The resultant depth grid is stored in the .flt format while the boundary dataset is store in ESRI's Shapefile format for use with geospatial software.

Version 4.1 of HEC-RAS includes the following new features:

1. New RAS Mapper Floodplain Delineation Capabilities
2. Hydrologic Routing Reaches Within an Unsteady Flow Model Run
3. New Flow Data And Boundary Conditions Editor for Unsteady Flow
4. Contraction/Expansion losses for Unsteady Flow
5. Minor Losses for Unsteady Flow
6. New Junction Hydraulics Option for Unsteady Flow
7. Groundwater Leakage for Storage Areas
8. Water Quality Modeling Enhancements
9. Sediment Transport Modeling Enhancements
10. User's Manuals and Help System

ArcView

HEC-RAS can also be used in the GIS programme, ArcView (With 3-D analyst and spatial analyst extensions). ArcView can read-in AutoCAD drawings for backgrounds and can create a DTM. In a similar way to river CAD, GEO-RAS can be used to create cross sections and other geometrical data for use in HEC-RAS and can be used to export water surface data from HEC-RAS back into ArcView to create flood maps with depths and extents of flooding.

AutoCAD

HEC-RAS can be used along with AutoCAD. A digital terrain model (DTM) can be constructed in AutoCAD and sections can be cut from the DTM and imported into HEC-RAS.

RiverCAD

RiverCAD by BOSS International Inc. supports the latest version of HEC-RAS and HEC-2 and is a standalone AutoCAD compatible application.

Relevant Journals

Mathematics and Computers in Simulation
Agricultural system
Water Resources Management
Water Resources Research
Agricultural Water Management
Journal of Water Resource and Protection
Irrigation and Drainage systems
Irrigation & Drainage Systems Engineering
International Journal of Engineering
International Journal of Computer Applications in Engineering Sciences
International Journal of Research And Innovations in Earth Science
International Journal of Simulation: Systems, Science and Technology
International Journal of Water Resources Development
Computing in Science & Engineering
Trans. ASAE
Trans. ASCE
Advances in Water Resources (Elsevier)
Irrigation Science

Questions

1. What is a model? Explain the need of model.
2. What are the considerations during development of a model?
3. Write short note on:
   Model calibration, validation, evaluation; Sensitivity analysis, implicit vs explicit method, finite element vs finite difference method
4. Explain different statistical parameters for model evaluation
5. Briefly describe the following models (with respect to capabilities, input, output, process/mechanisms involved, governing equations, and limitations): CROPWAT, UPFLOW, HEC-RAS, WEAP, SWAT, HYDRUS-1D, HYDRUS-2D

REFERENCES

Ali, M. H. & Abustan, I. (2014). A new novel index for evaluating model performance. *J. of Natural Resour. and Dev.*, *04*, 1-9.

Clarke, D. (1998). CropWat for Windows: User Guide. FAO, Rome, p.23.

De Laat, P. J. M. (1980). Model for unsaturated flow above a shallow water-table. Applied to a regional sub-surface flow problem. PUDOC, Doctoral thesis, Wageningen, The Netherlands., 126 p.

De Laat, P. J. M. (1995). Design and operation of a subsurface irrigation scheme with MUST. In Pereira, L.S., B.J. van den Broek, P. Kabat and R.G. Allen (Editors). Cropwater-simulation models in practice. Wageningen Pers, The Netherlands., pp: 123-140.

Donnigan, A. S. (1983). Model predictions vs. field observations: The model validation/testing process. p. 151-171. In: R. L. Swannand A. Eschenroder (ed.). Fate of chemicals in the Environment. Am. Chem. Soc. Symp. Ser. 225. ACS, Washington, D.C.

EPA, (2016). Storm Water Management Model (SWMM). URL: http://www.epa.gov/water-research/storm-water-management-model-swmm (accessed on: 27.01.2016).

Feddes, R. A., Kowalik, P. J. & Zaradny, H. (1978). Simulation of field water use and crop yield. Simulation Monographs. PUDOC, Wageningen, The Netherlands., 189 p.

Fox, M. S. (1981). An organizational view of distributed systems. IEEE Transact. *Systems, Man Cybernet.*, *11*, 70 – 80.

Loague, K. & Green, R. E. (1991). Statistical and graphical methods for evaluating solute transport models: Overview and application. *J. Contam. Hydrol.*, *7*, 51 – 73.

Mapp, H. P. & Eidman, V. R. (1978). Simulation of soil-water- crop yield systems: The potential for economic analyses. *South J. of Agric. Econ.*, *7*(1), 47-53.

Raes, D. (2002). UPFLOW – Water movement in a soil profile from a shallow water table to the top soil (capillary rise). Reference Manual Version 2.1, Dept. of Land Management, K.U. Leuven University, Belgium.

Raes, D. & Deproost, P. (2003). Model to assess water movement from a shallow water table to the root zone. *Agric. Water Manage*. *62*, 79-91.

Willmott, C. J., Ackleson, S. G., Davis, R. E., Feddeema, J. J., Klink, K. M., Legates, D. R., O'Connell, J. & Rowe, C. M. (1985). Statistics for the evaluation and comparison of models. *J. Geophys. Res.*, *90*(C5), 8995-9005.

Willmott, C. J., Robeson, S. M. & Matsuura, K. (2011). A refined index of model performance. *Int. J. Climatol.*, DOI: 10.1002/joc.2419.

AUTHOR'S CONTACT INFORMATION

Dr. Hossain Ali Md.
Principal Scientific Officer, and Head,
Agricultural Engineering Division
Bangladesh Institute of Nuclear Agriculture (BINA)
Bangladesh
hossain.ali.bina@gmail.com, mha_bina@yahoo.com

INDEX

#

A

B

C

D

G

H

I

J

K

L

M

N

O

Q

R

S

T

U

V

W

Y

Z